NATURALISM AND THE HUMAN CONDITION

Naturalism and the Human Condition: Against Scientism is a clear and compelling exploration of why naturalism, or what is otherwise known as the "scientific world-view," misrepresents what we really are as human beings.

Frederick Olafson offers up alternative ways of thinking about the human condition that have long been unrecognized by naturalists. Avoiding the metaphysical, he presses us to take a closer look at our human sense of being in the world, and shows that in its attempt to investigate human consciousness and intelligence, naturalism can only be understood through the very conceptual models that it rejects. Tracing the history of naturalism and its opponents, and drawing on a wide range of sources, including Heidegger, Merleau-Ponty, Husserl and Sartre, Olafson skillfully exposes the problems inherent in naturalism and raises some vital questions. The central chapters of the book discuss the implications of this on three main areas of the philosophy of mind: perception, language and brain-function.

Naturalism and the Human Condition is a vital contribution to the naturalist debate, showing that the way in which we understand ourselves as human beings has real consequences for all of us. It will be essential reading for all students of the philosophy of mind and language, and anyone interested in the debates about naturalism and scientism.

Frederick A. Olafson is Professor Emeritus of Philosophy at the University of California, San Diego. He is the author of *Heidegger and the Philosophy of Mind* and *Heidegger and the Ground of Ethics*.

NATURALISM AND THE HUMAN CONDITION

Against scientism

Frederick A. Olafson

London and New York

First published 2001
by Routledge
11 New Fetter Lane, London EC4P 4EE

Simultaneously published in the USA and Canada
by Routledge
29 West 35th Street, New York, NY 10001

Routledge is an imprint of the Taylor & Francis Group

© 2001 Frederick A. Olafson

Typeset in Times by The Running Head Limited, Cambridge
Printed and bound in Great Britain by TJ International Ltd, Padstow, Cornwall

All rights reserved. No part of this book may be reprinted
or reproduced or utilized in any form or by any electronic,
mechanical, or other means, now known or hereafter
invented, including photocopying and recording, or in any
information storage or retrieval system, without permission in
writing from the publishers.

British Library Cataloguing in Publication Data
A catalogue record for this book is available from the British Library

Library of Congress Cataloging in Publication Data
Olafson, Frederick A.
Naturalism and the human condition: against scientism/Frederick A. Olafson.
p. cm.
Includes index.
1. Philosophical anthropology. 2. Philosophy of mind. 3. Naturalism. I. Title.

BD450.O43 2001
146—dc21   2001019301

0-415-25259-8 (hbk)
0-415-25260-1 (pbk)

FOR CHANTAL

CONTENTS

	Preface	ix
1	Naturalism in historical perspective	1
2	Naturalism, dualism, and the natural attitude	10
3	The rejection of the given and the eclipse of presence	24
4	The substitution of language for presence: or words as things	46
5	What does the brain do?	65
6	Human being as the place of truth	84
	Conclusion	105
	Notes	109
	Index	113

PREFACE

This book is my attempt to justify to myself and to any one else who may be interested the hunch that has informed most of my work in philosophy.[1] The hunch is that, in a quite fundamental way, naturalism gets things wrong. In saying this, I am not making the familiar charge that it undermines religion and morality, although that may well be true. My thesis is rather that naturalism is confused and mistaken in the principal philosophical claims it makes. Since naturalism is once again in the ascendant, at least among philosophers, this seems a good time to lay out what is wrong with it.

The philosophical resources on which I draw for this purpose come mainly from the literature of phenomenology and especially existential phenomenology, but I try to avoid using the technical jargon of this or any other school of thought any more than I have to. The line of argument developed in this book is, however, heavily derivative from what I have learned from such thinkers as Martin Heidegger and Maurice Merleau-Ponty and others as diverse as Edmund Husserl and Jean-Paul Sartre. Since I am not interpreting their texts but reconstructing their arguments and adapting them to an intellectual situation they never had to deal with directly, I have not felt the need to make an express acknowledgment whenever I draw on the work of one or another of these thinkers. Nor have I felt any obligation to conform my views to those of any of these philosophers when they seemed problematic to me in one way or another.

In one respect, the way I propose to go about this task may appear to be open to question. There are many philosophers and scientists who have contributed to contemporary naturalism and correspondingly many differences in the way they argue their case. It is, therefore, important to make it clear that this book is not a study of their work or of the literature that has grown out of it over the past several decades. Instead, I have tried to pull out of it a number of large themes and strategies of argument; and these essays are best regarded as independent critical reflections on those themes. In proceeding in this way, I have not wanted to evade a general responsibility to characterize naturalism fairly and accurately; and I am prepared to be judged by how well I have succeeded in that regard. I have, however, tried to avoid tying

PREFACE

the case I make to any particular formulation of naturalism by an individual philosopher. Instead, I try to present a broad conception of contemporary naturalism that will, I hope, accurately state the central issues that it poses, but may not represent any individual thinker's views as accurately as he would wish.

I have tried to write these essays in a way that would make them accessible to the educated general reader. Accordingly, my effort has been to make my argument as compact and as continuous as I could; and I have avoided excursions into current controversies. If this procedure has given the line of thought I am expounding a more dogmatic aspect than I would wish it to have, I can only hope that the reader will understand that my intention has been to present an argument that stands on its own legs and that is as straightforward and cogent as I can make it. Consistently with this conception of my task, I have gone as light as possible on footnotes.

This is a philosophical book and it is mainly addressed to philosophical formulations of naturalism. At the same time, I hope that it can speak to the quite astonishing naivete by which the views of the scientific community in these matters seem to me to be characterized. In speaking of "naivete" here, I have in mind the belief that only antiscientific prejudice and a fuzzy-minded mysticism can stand in the way of a general acceptance of naturalism and the scientific world-view. This attitude properly belongs to scientism as the ideology of science rather than to science itself. Scientism in this sense is nowhere better manifested than in the recent book, *Consilience*, by a noted scientist, Professor E.O. Wilson.[2] The subtitle of this book is "The Unity of Knowledge," but what it defends is really the uniformity of knowledge. Professor Wilson identifies real knowledge with scientific knowledge; and he apparently thinks that the work of philosophy is to make surmises about what is as yet unknown but will presumably come to be known as science continues to advance. This pretty well excludes the possibility that philosophy might come up with arguments that could put a crimp in his vision of consilience.

He also says, in a formulation that sounds more promising than it proves to be, that the greatest intellectual issue facing us is how the humanities and the sciences are related to one another. It is quite clear, however, that, like C.P. Snow and most other commentators on this issue, he does not for a moment imagine that the scientific vision of the world and of human nature has anything to learn from any source whose authority is independent of science itself. My argument in this book is intended to show that the sciences of nature themselves have substantial presuppositions which cannot be formulated in their own conceptual mode. There are also ways other than natural science in which human thought can go forward from these same presuppositions without becoming, as Wilson seems to think they must, "transcendental" or otherworldly.[3] In this sense, an understanding of what human beings are that is not confined to the account that the natural sciences give of them is what the humanities try to convey when they are not, as they presently seem to be, intimidated by the aura of omniscience that hangs about science as the presumptive "theory of everything."

There is one other point that needs to be made here. I have spoken of philosophers and scientists and their views of human nature. This may make it sound as though what is at issue here were some arcane dispute within the academy in which no wider public has any real stake. In my view, that would be a serious mistake; and I will try to show that the way our identity as human beings is understood has real consequences for all of us. It is true that even the most serious misconceptions of what is distinctively human cannot change what we are. What is possible, however, is for the implicit understanding we have of ourselves and the means we have of expressing it to suffer substantial erosion. This can happen when the distinctions that are marked in the language of everyday life come to be regarded as a kind of rough makeshift by comparison with the exactitude of the scientific language that will eventually replace them. The devaluation of all thinking in humanistic terms that typically results from this comparison can issue in an effective denaturing of all the concrete forms in which our humanity expresses itself. More generally, the exercise of our distinctive powers depends in some measure on their being actively cultivated; and their proper cultivation, in turn, requires that they be recognized and conceived in some adequate manner. That is an effort to which this book seeks to make a contribution.

1

NATURALISM IN HISTORICAL PERSPECTIVE

I

The idea of nature has figured prominently in the efforts human beings have made to understand themselves. Sometimes it has done so as a conception of what we ourselves really are and, sometimes, of what we are not. In a kind of compromise between these two views, "nature" has also been conceived as an ideal to which our lives are to be conformed. This was the teaching of the Stoics, but, as John Stuart Mill pointed out, it seems to postulate, first, a distinction between nature and human nature and then makes the former a norm for the latter.[1] Inevitably, one wonders why, if there is this difference between the two, the human should conform itself to the non-human. If, on the other hand, we really are part of nature, how does it come about that we can (and often do) live "un-naturally"? May this "failure" on our part not be susceptible of another interpretation? Perhaps we are not just unusually perverse members of the natural order, but rather, in certain significant respects, do not belong there at all. There is, evidently, a prior question here about where, in the wider scheme of things, we do belong. That is also a question I want to address in this book.

Whether we are a part of nature or not depends, of course, on how the idea of nature is itself construed. Unfortunately, precision and stability of meaning have not been very common in discussions of this topic. The nature of the Stoics was a highly moralized affair and by no means like the nature that was described in the last century as being "red in tooth and claw." Then there is the nature that is now conceived in terms of molecules and atoms and sub-atomic particles. And even before science or philosophy made its appearance on the historical scene, the place of human beings in the world had been the theme of myth. There were stories in which the earth was the mother from whom we are all descended; and others in which a male sky god appears to have represented something like a transcendence of these earthy beginnings. But when it comes to delimiting the concept of nature, the most crucial issue has to do with the status of non-human animals. They are almost always held to belong to nature even if human beings do not. But if the basis for this distinction was once thought to be unproblematic, that is certainly no longer the case. Even so, there are still many people who would be indignant at the suggestion that human beings are part of the natural order in just the same way as are "the beasts of the field."

In the history of thinking about these matters, the intellectual and moral powers of human beings have been held to distinguish us from other living creatures. Even today, when nature is typically understood as a system of material entities and processes that is governed by invariant laws and indifferent to human purposes and values, it is evidently these powers that enable us to conceive nature as just such a prior, independent order and thereby, arguably, give us a special status. Another reason for asserting the distinctness of human beings from the natural world has been the fact that although we are obviously dependent in countless ways on the nature in which our lives are embedded and against which we are often helpless to defend ourselves, we are also able to intervene in natural processes and harness them to our own purposes to a much greater degree than other animals are. This fact about us may even have inspired, at least in part, the view that represents human beings as a special creation—in his "image and likeness"—of a God who is certainly not part of nature and is supposed to have set man over nature as a kind of ruler.

The idea that human beings are in nature as its rulers and that their true affinities are with a quite different sphere of reality has also had its critics. Some of these have been motivated by a perception of the mess we appear to have made of our supposed stewardship over the earth. Both before and after Rousseau, there have been thinkers who have taken a more skeptical view of the great work of civilization that human beings have claimed to be carrying out by intervening in the natural order. Such critics have reserved their admiration for those societies that have changed their environments only minimally. The broad implication of this kind of critique is that we humans should accept a quite modest place within nature and stop deluding ourselves with thoughts of some exceptional vocation that has been assigned to us.

Nevertheless, the main philosophical tradition in the West has persisted in defending the idea of such an exceptional status for human beings. As already noted, the special character of the human mind, especially as it has been interpreted in the thought of such modern philosophers as Descartes and Kant, as well as Hegel and Husserl, has been thought to justify that status. As the ultimate ground of human culture in all its forms, including the sciences of nature, mind has appeared to be proof against all attempts to reduce it to the status of a by-product of natural processes. Even when the wider syntheses proposed by these philosophers have gone down before critical attacks, the privileged position of "consciousness"—the term most frequently used to designate the defining function of the human mind—has typically survived even though its relation to the natural order has remained anomalous.

In our century, however, a quite different line of thought has developed that questions the idea that human beings cannot be assigned the status of creatures of the natural world. In a sense, it is linked to the very technologies that have made it possible for us to turn the natural world into our workshop to a far greater extent than ever before in human history. If these technologies have given us our sense of being the masters of, if not the universe, at least the natural world in our neighborhood, they also contribute to another quite different conception of our nature and status. These technologies are now closely linked with what are called the sciences of nature; and the methods of inquiry of the latter have achieved the status of paradigms of knowledge generally. By comparison, the kind of knowledge that human beings have traditionally accumulated of their own "natures," as well as the moral and teleological terms in which this knowledge has typically been cast, have come to seem both fragmentary and feeble. Inevitably, the idea took shape that "real" inquiry into the controlling regularities of human life must proceed in accordance with the methodologies of the natural sciences. The progress that was made in medicine and in the physiology of the human body offered strong confirmation of this line of thought. It has gradually been extended into other areas that had been thought to be the province of the soul or the mind and thus not physical or "natural" in the way that, say, the circulation of the blood is. As a result we now live in a world in which systematic inquiries into our social and mental/emotional nature have achieved widespread acceptance and are brought under the rubric of "science" as "social" and "human" sciences.

The paradox inherent in these developments has not passed unnoticed. All of our efforts to make ourselves masters of our situation in the natural world have proceeded under the auspices of what is still commonly referred to as the Enlightenment. This was the movement of critical thought in the eighteenth century that, on the basis of the scientific revolution of the preceding century, undertook a searching critique of the religious and metaphysical underpinnings of the Christian civilization of Europe. The great natural philosophers of the seventeenth century had shown that nature can be conceived in mathematical terms that altogether bypass the teleological ordering envisaged by Aristotle. The idea of the Enlightenment *philosophes* was that scientific methods of comparable exactitude could be brought to bear on human life and on human institutions and practices that had traditionally been protected against critical scrutiny by the veil of Christian myth. The comprehensive program of demystification and secularization thus initiated was to become the agenda of liberal political movements throughout the Western world and beyond it as well.

The paradox generated by this conception, which was to shape the modern world in very significant ways, consists in the fact that the methods of inquiry that had been designed to deal with inanimate nature—that is, with *things*—were to be applied to the human beings who were the authors of this project.[2]

This is what has been called the "dialectic of enlightenment" by which people are denied any status that sets them apart from the natural world that they have already reduced to a system of objects. The implications of this new status, especially for the ethical dimension of our relation to one another, are obviously very considerable. But once this great engine of scientific inquiry and rational administration had been set in motion, it proved almost impossible to stop it or to modify its procedures even when the fit between them and the situations to which they are applied becomes more than a little problematic.

Not surprisingly, there has been a reaction to the vision of human life thus generated. Since Max Weber, it has become customary to speak of the "disenchantment" of the world that these rationalizing procedures have brought about. What this term expresses is a sense that the interest—most notably, the appeal to the imagination—of the world as so conceived has been radically diminished. It is certainly true that a world in which we are not the masters, even in principle, and are instead at the mercy of beings and powers over which we have little or no control, is likely to be more interesting, often in a quite scary way, than one in which everything is held to be explainable and predictable. The old saying that Vico quotes—*homo non intelligendo fit omnia* (freely: by virtue of not understanding the world around him, man himself becomes the model on which all things are interpreted)—reminds us of a time when human beings had only themselves and, more specifically, their purposes and the fears and hopes associated with them as a model on which to interpret the world around them. In those circumstances, it was not arrogance or pride, but the lack of an alternative that led people to read these same hopes and fears into a natural world that we now think of as being utterly indifferent to them. We also know what an extraordinary corpus of stories and images the Greeks and other peoples elaborated in the state of *Urdummheit* that supposedly preceded the emergence of a rational and ultimately scientific world-view. It is hard not to wish that a walk in the woods could be as magical an experience for us as it evidently was for them. But, in the words of W.B. Yeats, "the woods of Arcady are dead and over is their antique joy"; or so it is now widely and regretfully felt.

II

The philosophical controversies to which all these developments have given rise set the context for the theme of this study. That theme is naturalism—a movement of thought that not only takes its name from "nature" but assigns an unqualifiedly positive valence to the fact of our being part of nature. In the tradition of the Enlightenment from which it is itself descended, naturalism was originally a reaction against religious ideas of a supernatural domain to which human beings were supposed to be somehow akin. It was also directed against philosophical systems like idealism that were thought to have much too strong an affinity with a religious view of the world and to do scant justice to the role in our lives of the natural world into which we are born.

Naturalism as a movement of opposition to these systems of belief also came to be associated, in the thought of a philosopher like John Dewey, with the kind of social liberalism that sought, early in this century, to open up new horizons of thought and practice. Religious and social orthodoxies, by contrast, were identified with constraints that in one way or another block a full development of natural human powers and potentialities for self-realization. As a result, naturalism came to be understood as a liberation from the dogmas of supernaturalism and the conservative social order for which they served as an ideology, as well as a declaration of independence for scientific inquiry into the nature of the world and also into human nature.

Beyond these very general affinities, however, this kind of naturalism rarely committed itself to philosophical theses that characterized nature in a way that really made clear what it could and could not comprise. Certainly, it would be hard to gather from Dewey's writings, for all their pro-science orientation, whether the ontology or conception of reality espoused by the natural sciences was implicit in naturalism as he understood it, and if not, why not. In any case, after Dewey's death in 1952, the appeal to philosophers of the term "naturalism," as well as the appeal of the rhetoric associated with it, appeared to fall off quite substantially. This was certainly not because the supernatural or the spiritual held some new attraction for them. The reason is more likely to have been that the expanding role, within philosophy, of logic and the logical analysis of language tended to turn philosophers away from an interest in the "nature of things" in the old comprehensive sense. That, it came to be thought, was someone else's business. Although this "someone else" was generally held to be the natural scientist, the work of philosophy was conceived as the clarification of the conceptual and logical preconditions for anything we might claim to know about the world. As such, it was supposed to be largely independent of the work of the various scientific disciplines.

More recently, however, naturalism has reappeared in the purlieus of American philosophy and it has done so in a new guise. Philosophers have been trying to make common cause with science for some time now, perhaps on the policy of "If you can't beat them, join them." As a result, they no longer show as much interest in abstaining from first-order pronouncements about the nature of things as they did when language was their principal theme. It is not uncommon now to hear philosophers describe their work as being not so much preliminary to, as more or less directly continuous with, that of the natural sciences. With this sense of a direct convergence of purpose between science and philosophy, there has come a revival of the slogans of naturalism. The older naturalism certainly identified itself with science and the scientific world-view; but, as has already been noted, the terms of its allegiance were rather vague and had more to do with using science to put other competing world-views out of business than it did with any precise ontological commitment. Now, however, these rival positions are pretty well defunct and their philosophical theses no longer command real interest, so a struggle against them would be unlikely to stir much interest.

What is of great significance for this study is that, during this hiatus in the public sponsorship of naturalism, its philosophical promoters have, for reasons that will need examining, reformulated naturalism as an all-embracing materialism or "physicalism," as it is now more often called. There is no longer any room for saving ambiguities like the concepts of "experience" and the "psycho-physical" that were still coin of the realm in the older naturalism. These were part of the residue of the empiricism from which it had not yet clearly distinguished itself. As a result, for a contemporary naturalist the only conceptual system in terms of which the world-process can be reliably characterized is held to be that of the physical sciences of nature. On such a view, the world and nature are one and the same and everything in them is of the same ontological type.

The kind of naturalism that has developed in the last few decades is significantly different in a number of respects from the classical materialism of a Democritus or a Hobbes. Democritus simply asserted that everything was particulate matter moving in the void and Hobbes was equally direct in the claims he made about the physical character of mental processes like sensation. This kind of flat-footed naturalism now seems crude to most philosophers. They have had to fight battles against other philosophical traditions that stem from Descartes and Kant and they are incomparably more sophisticated than their predecessors when it comes to working out strategies of argument to defeat the claims of these and other non-naturalistic thinkers. In all of this, a philosophical interest in language, informed in good part by the work of Ludwig Wittgenstein and others, has played a major role.

Even with all these new currents of thought, the rhetoric of this renewal of naturalism remains pretty much the same as it was in the past. The message is still that we humans are part of nature—creatures of the natural world, through and through. The trouble with this thesis has always been not so much what it affirmed—the obvious similarities between human beings and other denizens of the sublunary world—as what it was intended to deny. This was the idea that there could be any deep ontological differences among the kinds of entities, ourselves included, that in one way or another, can be said to be "in the world." Every half-way emphatic suggestion to that effect has been treated by the proponents of naturalism as a reversion to the religious ideas of an otherworldly kind that they most wanted to discredit. Although that goal would not seem to require that everything that is in some sense a part of the natural world be in it in just the same way, that is what naturalism does require when its hostility to anything that smacks of the supernatural evolves into an express ontological requirement that our lives and our actions and we ourselves be conceived in physicalistic terms.

It is this development within naturalism that has transformed it from a rather vague rhetorical stance into an uncompromising thesis that has become widely influential within American philosophy. It is hardly too much to say that in this guise naturalism has become a kind of all-purpose philosophy for those who think the natural sciences can provide answers to all the problems with which philosophers have traditionally tormented themselves or at least to the ones that do not simply dissolve on closer inspection. So convinced of its truth are the sponsors of this thesis that they have great difficulty crediting the idea that there is any alternative to it that deserves to be taken seriously and is not just a shameless reversion to the superstitions of the past.

I will refer to this position as "hard" naturalism and to the older kind whose principal interest was in anti-supernaturalism as "soft" naturalism. (When the term used is simply "naturalism," it will denote hard naturalism unless otherwise indicated.) The broader import of the movement from one to the other is still not well understood even by many of its proponents and certainly not by the wider educated public that might be expected to take an interest in such developments. As a result, the assumption remains largely undisturbed that naturalism still constitutes a principal element in the intellectual armory of thoughtful people of progressive and liberal temperament. We tend—rather wishfully—to assume that that other set of intellectual and moral interests that we call "the humanities" can co-exist with this radicalization of naturalism as the ultimate set of postulates in terms of which a "theory of everything" is to be formulated. We have been assured, after all, by C.P. Snow that the party of science has "the future in its bones." How, then, can there even be thought to be any serious alternatives to naturalism as the philosophical standpoint from which we are to understand ourselves and our world?

This book will challenge all these assumptions. It will do so by showing, first, that the argumentation by which naturalism in its physicalistic incarnation has tried to demonstrate its validity is faulty. Not even the prestige of "Science," I will claim, can carry the day for philosophical theses that are as deeply and perversely paradoxical as those of this transformed naturalism prove to be on closer inspection. But, beyond this critique of naturalism, I will show that there are alternatives for thinking about the human condition other than those that are recognized by naturalists. In doing so, I will resist the established presumption that any critique of naturalism must try to reach out beyond the limits of anything we can put to an empirical test and postulate a form of reality that is, in the fallen sense of that much put-upon word, "metaphysical." Instead, my claim will be that, instead of venturing into unknown realms, we need to take a second and closer look at certain matters that are so familiar to us that we tend to think we already understand them as well as anyone possibly could. In a way, it is true that we have such an understanding, but it is largely inarticulate and certainly does not find its way into the theories of the world and of human nature that dominate our intellectual life. Getting at that understanding and the ways in which it has been overlaid both by naturalism and earlier by the conceptions of the mind that naturalism attacked will be a major task of this book. If that can be done, it will also become clear how different the world we live in and we ourselves are from the standard accounts that both science and philosophy have given of such matters.

In drawing a distinction between the "soft" and the "hard" versions of naturalism and addressing this study to the latter rather than to the former I do not want to be understood as implying that the philosophical or historical significance of the former is negligible. Almost everyone now agrees that our lives are deeply embedded in the natural world, even though we may not know exactly how this works, but it was not always so. The freedom of the relevant sciences to investigate everything about us that is in any way connected to or dependent upon processes that occur in the natural world had to be won through a hard struggle against stubborn and powerful opposition. And if, in the Western world at least, the battle to secure that freedom has been pretty well won for some time now, there can be no assurance that it will stay "won" forever or that it will prevail throughout the world as it has in the West.

To say this is not to concede anything to the scientific world-view that may have to be contested at some later point. It is, after all, one thing to demand the most complete freedom for every branch and kind of inquiry to press its claims as far as it can, and quite another to endorse any special claim that may be made by those who engage in this effort. Accordingly, the logic and methodology of the natural sciences should not be assigned any antecedently privileged position that rules out in advance any thought of an alternative to them and makes any deviation from them automatically suspect. Indeed, it may well turn out that the application of the word "science" to certain disciplines of thought which thereby become "social" and, more recently, "human" sciences will prove, on closer scrutiny, to be seriously misguided. The recent controversy about "sociobiology" is an example of how such extensions of the assumptions of one branch of knowledge to another can run into principled opposition that is not motivated by any desire to protect a certain terrain of inquiry from searching and rigorous investigation. But even the severest critic of sociobiology would presumably not contest the right of a biologist to develop analogies between, say, insect and human societies. The contrasts that may come to light between the terms of such analogies can, after all, prove to be as suggestive as the supposed affinities they postulate.

There is one further point that needs to be made here. What has just been said does not mean that soft naturalism is immune from criticism. It is, in fact, vulnerable to criticism in one very significant respect. It has regularly finessed the question of its relationship to its "hard" sibling—physicalism. The reason for this appears to have been that once the allegedly otherworldly alternatives to naturalism had been widely discredited, the exponents of soft naturalism did not feel any strong interest in what might be called the fine structure of the natural world in which, they claimed, everything took place. Although, as stated, they were typically not committed to any explicit form of materialism, their fear of a recrudescence of supernaturalism led them to evince a strong antipathy to ontological issues as such and an equally strong predilection for the idea of nature as a great cosmic pudding into which everything will eventually be folded back. It was as though all the really important philosophical issues had been taken care of once the main philosophical tradition had been shown to be driven by a "quest for certainty" that could achieve its goal only by postulating some non-natural form of reality. And yet it is really quite clear that such issues do remain and that chief among them is precisely this question about the unique authority of the natural sciences to determine what there is in the world.

2

NATURALISM, DUALISM, AND THE NATURAL ATTITUDE

I

There can be little doubt that for most people (and even for many philosophers) the strongest argument for naturalism is that it is supposed to be the scientific approach to a variety of issues that have been thought of as philosophical but that philosophers do not seem to have had much success with. As a result of this prior commitment to the superiority of scientific inquiry, it appears to have been felt that little argument was needed to justify naturalism beyond showing how hopelessly inept the theories traditionally put forward by philosophers really are. On this view, scientific inquiry into human consciousness and intelligence owes essentially nothing to the philosophical treatment of these subjects that preceded it. In actual fact, however, it is impossible to understand naturalism otherwise than as a development out of and reaction to the very kind of dualistic theory of mind that it so strenuously rejects. It will also be a major thesis of this book that the resulting linkage is still in place.

If the matters at issue in this discussion—initiated by naturalism's effort to set aside specifically philosophical theories of mind—are to be understood, something needs to be said first about the way the concept of mind itself has developed. Its origins are surely to be found in the concept of the soul and of human nature as a compound of body and soul. But if mind was initially a part or an aspect of the soul, the cognitive functions of the soul were eventually distinguished from others that had to do with life and motion; and it was with the former that mind came to be principally identified. Although minds have been variously conceived, their differences from bodies have always been strongly emphasized and in many accounts they have been held not to be physical entities at all. The philosophical elaboration of the resulting concept led to the mind's being understood as the metaphorical place where ideas or "representations" of things in the world are housed and where characteristically mental operations are performed upon them. These operations consisted largely in judgments in which the character of things in the extramental world was defined. The business of the mind, in other words, was knowledge; and although it included the mind's knowledge of itself, this knowledge was mainly concerned with things in the world which, in this idiom, had to be thought of as "external"—that is, outside the mind understood as a particular entity of a very special kind.

The real point about all this is that the concept of the mind has a constructive character. This means that it is not so much a direct transcription of something we can be said to "experience," as it is a way of ordering in thought the heterogeneous elements that turn up in our commerce with the world and with ourselves. As such an ordering, "the mind" cannot claim to be insulated from critical commentary on its design or revisionary suggestions that would replace it with some alternative. Nevertheless, it has to be acknowledged that dualism as the ontological formulation of the contrast between mind and body has established itself so firmly that it is typically taken to have been laid down in the original blueprints for the kind of world in which we live. Its authority has been such that in the modern period the mind has come to be regarded as not just another entity or "subject" (in the old sense of that term) but as the "subject of subjects"—in effect, *the* subject in the modern sense of the term.

When it is so understood, the mind is responsible for setting up criteria of reality by which what is in the world can be separated out from what may appear to be so, but really is not. Interestingly, "subjective"—the adjective formed from "subject"—has been mostly identified with whatever turns out not to be properly assignable to things as they really are in themselves and thus forms a kind of residue in the mind. The contrasting term, "objective," again in its modern usage, designates whatever has turned out to be so assignable. Clearly, however, in order to be able to make this distinction, the mind must be as familiar with what turns out to be objective as it is with what is found to be merely subjective. This means that the mind must contain representations of both. This is a point of the very greatest importance. It is expressed here in the idiom of dualism, but it is susceptible of another formulation that dispenses with all mental intermediaries. As so formulated, it will be central to an understanding of the concept of being-in-the-world that will be substituted for that of mind at a later point in my argument. Historically, however, the mind—the subject—has been identified with the subjective; and by providing an asylum in which properties that did not qualify as objective could be accommodated, the mind formed an important part of what has been called the "metaphysics of modern science."[1]

The idea that the mind's function was to represent things in the world rather than to encounter them directly grew out of insights into the fallibility of perception and, more generally, out of objections to uncritically realistic assumptions about cognition. Perception is, indeed, not free from error and sometimes it does seem to invite our belief in things that prove to be quite unreal. From this, it has been inferred that we should not just assume that whatever appears to be put before us in perception must exist. Then, too, because normal perception and its fraudulent counterpart have been thought to be indistinguishable when considered simply as initial presentations or appearances, some other criterion of reality was evidently required before any such appearances could be accepted as veridical. That criterion was generally held to be provided by the higher mental faculties and especially by thought in the service of reason. Outranked in this way, perception was assigned a relatively modest role in the search for truth.

Another weakness of common-sense realism was its failure to develop the implications of the fact that mental functions are mediated by physical processes occurring inside and outside our bodies. This kind of realism has been called "naive" by its critics since it seemed to claim that one had only to open one's eyes and the world would be there. This lacuna in the common-sense view plainly needed to be filled and dualism provided an account that at least seemed to do just that. It distinguished between the qualities in experience that really do have counterparts in the world—these were spatial and quantitative in character and were often called "primary"—and those "secondary" qualities that had no such counterparts and were produced in us by the action of things that were the bearers of the primary qualities. In this way dualism replaced the realistic view with one in which the perceptual relation of the mind to its object is mediated by processes in the body although the terms in which these were conceived were still very crude. On this view, perception occurs in a human being whose sense organs have been stimulated; and it does so at the end of a process of transmission outside and then inside the body. Originally, perception together with the sense qualities it comprises was held to take place in the mind rather than, as later came to be thought, somewhere within the physical confines of the body. But, in either case, something that occurs at the end of such a process can hardly be identical with the object by which that process is set in motion. The resulting conception of perception was thus both dualistic and representational. The content of perception was identified with a state of the mind and one that was supposed to represent an "external" object which itself never directly appears.

Although this kind of dualism became the standard and orthodox account of perception and, derivatively, of all other distinctively mental functions, it was not long before the very serious paradoxes it generates came to light. One main source of these was the fact that a concept of the mind as a kind of container in which ideas and representations are sequestered has the effect of separating perception from its object in a way that was to create just as many problems as the "naivete" of the common-sense account of perception had. The classical puzzle to which a dualistic view of perception gives rise is whether there can be any reason to think that a perception, understood as the final outcome of a process of neural transmission, in any way resembles the "external" object that it is supposed to represent. (One might also ask what reason the person in question would have for viewing the appearance that is the outcome of this process as a representation at all.) Plainly, there can be no empirical confirmation (or disconfirmation) of any such resemblance since any perception that might claim to confirm it would itself be open to the same question. That dualism both gives rise to this question and makes it impossible to answer it must be viewed as a grave incoherence in that theory.

These criticisms of dualism are philosophical in character and they are mainly motivated by epistemological considerations. Although they cut very deep, it is important to remember that in the writings of Berkeley and Hume, for example, they never called into question the reality of mental states or of what we are now more likely to call "experience" or "consciousness." Indeed, as the example of Berkeley shows, their tendency was not to make anything like materialism more plausible, but rather to provide arguments for an idealistic view of reality as being somehow "mental" or spiritual in its overall character.

This was to change, however, when such criticism began increasingly to reflect the criteria of reality that are implicit in the methods of the natural sciences. Such a development was hardly surprising inasmuch as the status of mind as it is envisaged by dualism was, in spite of its utility as a place where the purely subjective elements in experience could be domiciled, utterly different from that of the objects these sciences themselves dealt with. It became increasingly evident that if the epistemic criteria employed in the study of natural phenomena were applied to the study of human beings (and especially human minds), the latter would be hard pressed to meet them. More concretely, every knowledge claim we make about ourselves as minds would have to be susceptible of being put to the test through some form of repeatable observation, and thereby confirmed or disconfirmed by the members of an open community of inquirers.

As a result of these developments, it would hardly be too much to say that the problem of mind became the problem of how one could avoid having to acknowledge the reality of something as oddly circumstanced in the natural world as the mind was supposed to be. The difficulty has been how to deal with the unhappy legacy of dualism as the philosophical co-adjutor of the new mathematical concept of nature. That legacy was a great accumulation of mental debris that was made up of all those features of experience that at first seemed to have a legitimate place in the architecture of the world, but turned out not to qualify as properly "objective." One way to go about this is simply to consign all the non-standard odds and ends with which the mind had been filled to non-being, as the capacious domain of what is not really the case. These were, one might argue, just so many things that had been mistakenly thought to be real but proved not to be. Since what is mistakenly thought to exist need not take up any space anywhere, it should not be necessary to keep the mind going as a kind of shelter for ontological unemployables.

Popular as this resolution of the puzzle remains to this day, it is of a piece with the general rationalistic effort to assimilate perception to thought. More pertinently, it simply will not work. Colors, for example, are not just qualities that we mistakenly think things have. If they were, it would be hard to explain the fact that we are all apparently making the same mistake and keep on making it even after we have ostensibly been convinced by philosophical and scientific arguments that it is a mistake. Something, evidently, is really colored; and if that something is not a piece of the outside world, it must, by the logic of dualism, be something in the mind that does not qualify for inclusion in the official constitution of the world. But if varicolored sense data are in the mind, they can hardly be there in isolation from the representations of physical properties they are associated with. As has already been pointed out, unless the properties that turn out to be objective were somehow present "in the mind" to begin with, we would be unable to acknowledge their independent status. Thus, if a color is in the mind as a colored expanse, then that phenomenal surface will have to be there as well and so on and on. In the end, a complete world will be there—a world that includes everything in the "external" world and a number of other allegedly non-objective properties as well. It is certainly hard to see how it can make sense to claim that this vast assemblage of properties—both objective and subjective—is "in us" or rather, in each case, "in me." It is not an exaggeration to say that the status of those properties that, according to the scientific view, do not belong in the world is still unresolved.

To sum up, dualism is a philosophical response to certain anomalies in our pre-philosophical and pre-critical understanding of ourselves and our place in the world. It postulates an entity called "the mind" that is a part of a human being but not of its body and performs various functions that are then designated as "mental." Matters are further complicated by the fact that the mind has the ontological status of a thing or substance. It has been suggested that this idea of the mind is a reworking of the generic thing–concept, by which it is turned outside in so that what were the properties of things outside the mind become representations of those properties in the mind as a new kind of interior. This conception acknowledges, after a fashion, the fact that we are also in the presence of something—a world—that is not ourselves; but it interprets this fact in its own way. The world, it holds, is represented within this notional entity—the mind—and its reality is acknowledged principally in thought and not in perception. But how representations that are internal to such an enclosure could ever disclose the existence of something that is *ex hypothesi* external to it, no one has ever been able to explain. What is apparent is that the mind conceived as this curious bit of conceptual architecture is precluded from sustaining a relation to anything but its own representations. It cannot, that is, be anything but a worldless subject.

I have described a process in which the dualistic concept of mind passed from a condition in which it was an essential adjunct to the modern scientific world-view to one in which it came to be regarded as an ontological solecism. Increasingly, as it became apparent that even what had been classified as purely subjective claimed a certain *droit de cité* for ontological purposes, scientific thought began to react adversely against anything as resistant to its own categories as mind has been held to be. What may be just as significant as this allergic reaction to mind–body dualism is the fact that defenders of the scientific world-view have typically assumed that the idea of such a non-physical component of human nature represents the only alternative to their conception of things that needs to be taken seriously. Critics of naturalism as well have often accepted this idea and have concluded that their opposition to naturalism commits them to a defense of the kind of dualism that naturalism rejects.

Naturalism emerges from what has been said thus far as a critique of another philosophical theory of human nature—the theory, in fact, to which we owe the concepts of the mind and of the mental as they have been understood in the modern period. It has to be said that this critique itself is not particularly original insofar as it relies on the kinds of epistemological arguments that were brought against dualism earlier in the modern period. More will be said about this in the next chapter. There is good reason to doubt whether a critique of dualism, however successful, can establish the truth of naturalism. For one thing, it looks as though, by predicating its argumentation on what is in fact a prior philosophical notion—the conception of mind as a private, inner domain that is inaccessible to, and unverifiable by, others—naturalism has failed to take an independent look at the datum for all these discussions and neglected the possibility that there may be alternatives to dualism that are not naturalistic. It simply assumes that whatever is described as being "given" or "present" must be "in the mind" in some objectionable dualistic sense of that expression. Naturalism, accordingly, rejects root and branch anything that is so described and it does so without any sense that it may thereby have cut the ground out from under its own familiarity with the world about which, after all, it has a great deal to say. Consequently, it opens itself to the suspicion that its ability to continue holding forth about the nature of things must depend on its tacitly drawing on resources—most notably, those afforded by occasions on which things in the world do show themselves—that it denies it has.

II

It should be evident by now that naturalism and dualism are not just rival theories but stand in complex relations of interdependence with one another. They are linked, in the first instance, because naturalism defined itself largely through its rejection of dualism. It thus begins its analyses with the concepts of mind and the mental already on hand and, with them, all the baggage of "inner states" and "introspection." It then proceeds on the assumption that if this mentalistic apparatus can be shown to be hopelessly at odds with the requirements of scientific method, the case for naturalism will have been made. Beyond this, there is an even deeper linkage between the two since naturalism is also the residuary legatee of the estate of dualism. In plain terms, naturalism inherits the body that dualism, on its demise, leaves behind. But if that body was itself the product of a questionable surgical procedure that reduces the human body to the condition of a natural object, its value to the heir or to anyone else may turn out to be quite problematic.

In Chapter 5 it will be shown that hard naturalism has never really succeeded in detaching itself from dualism as completely as it typically claims to have done. As a result of this imperfect separation, certain ambiguities arise as to what naturalism is saying about the mind and the body. These ambiguities cut both ways. By making it appear that naturalism makes some place for the "subjective" even as it reduces it to brain function, they make naturalism more credible and thus more attractive to some of its adherents. In doing so, however, they also create a distinction between an exoteric and an esoteric version of naturalism that may prove disillusioning to the faithful if they become aware of it. In any case, if the considerable complexities that characterize this relationship are to be sorted out, we will need to take another look at the starting point of this whole evolution.

Dualism has already been shown to be the outcome of an earlier critique of pre-philosophical beliefs that are slightingly referred to as "naive realism." Is it possible that the latter was not just a regime of error as it is usually thought to have been and that it may even offer a more satisfactory basis for thinking about ourselves than do its two more highly theoretical successors? If so, it will be necessary to show that the natural attitude is not simply the point of departure for this dialectical sequence, but can be formulated as a philosophical position and one that has not been hopelessly superseded by the critiques to which it has been subjected.

Admittedly, there is not much in the way of precedents for such an undertaking. There appears to be a deep unwillingness on the part of many philosophers to acknowledge the possibility that scientific knowledge may have its roots in anything akin to the kind of common-sense beliefs that usually have to be overturned if deep insights into the workings of things are to be achieved. One reason for this attitude is simply the fact that even a bare acknowledgment of this fact would have to be couched in language that would strike many as being itself extra-scientific and therefore in all likelihood un-scientific. This in turn reflects the idea that if science has any philosophical presuppositions, it should be possible for them to be stated in a language that is as close as possible to that of science itself. Naturalism is in fact itself just such an effort to substitute clear-headed scientific thinking for misty adumbrations of how the self may have emerged from its matrix in the natural world.

Among the philosophers who have not shied away from this task, a number of names stand out. It was Edmund Husserl who first spoke of the natural attitude as the deeply implicit and universal conviction that what we see and otherwise perceive are objects that lie for the most part outside our own bodies and/or minds and continue to exist with the properties we observe in them even when they are not being perceived by us or by anyone. At the same time, however, he argued that this attitude had to be suspended precisely because its commitment to the existence of the world went hand in hand with a deep obliviousness to the true nature of the consciousness that it exemplified itself. This suspension was supposed to allow the structures of pure consciousness to emerge into the kind of clarity that is precluded when we are so fully absorbed in our dealings with things in this same world that we think about knowledge and consciousness in terms that are systematically inappropriate to them. If the claim of the natural attitude was that the world is simply *given* to us, Husserl tried to show that on closer inspection a great deal more is involved in this givenness than the natural attitude acknowledges. The account he proceeded to give of how we have a world did not, however, completely succeed in dropping the mentalistic assumptions of the dualism he was trying to avoid. It has also never been altogether clear where that account leaves the main theses of the natural attitude itself about the existence of the world. Was the natural attitude supposed to be resumed in a more sophisticated and self-aware form or was it effectively abrogated?

Something very like the natural attitude was also implicit in the Scottish common-sense philosophy of the eighteenth century and in the more recent Oxonian philosophy of ordinary language that was so heavily indebted to the thought of Ludwig Wittgenstein. It cannot be said, however, that either of these did much to place the natural attitude on a secure philosophical foundation. Without a doubt, the most considerable contribution to such a statement of the natural attitude was made by Martin Heidegger. In his early writings, the thesis of the natural attitude becomes the fundamental fact that displaces all the intramental apparatus that had been designed to explain how human knowledge of such a world was possible. He did not, however, do much to invalidate the criticisms—noted above—that have been brought against realism, whether naive or otherwise. These are the criticisms that turn on the fact of perceptual illusion and on the role of the brain and the nervous system in our "mental" activities.

In what follows I want to give a brief sketch of a number of considerations that ought to qualify our automatic rejection of the natural attitude and make us look at it with new interest and just possibly respect. For example, when the natural attitude is accused of being too credulous vis-à-vis the presentations of sense, it should be remembered that it may be a strength and not a weakness that it does not go about dealing with these facts in what has until recently been the orthodox manner. It does not, in other words, devaluate perceptual experience by interpreting it as a tissue of sense presentations that are, as such, prior to any distinction between truth and falsity and take on a truth-value only through the intervention of our faculties of judgment.[2] That approach generates its own well-known difficulties; and there is good reason to think that there are better ways of dealing with the fact of error. It is after all a familiar experience that things can look otherwise than they really are; and there does not seem to be any reason why the normal range of perspectival variations about which we make no difficulty should not be extended to include those experiences of perceptual illusion about which we do.[3]

In the matter of the role of the brain and the central nervous system, it is true that the natural attitude has very little, if anything, to say. Here again, however, this very silence may be a kind of strength. The blunt fact is that the brain plays no part in our own experience of the various functions—of perception, of thought, of action—that are usually classified as being mental. And it should also be recalled that when dualism tried to set the mind in a causal relation to the brain, the result—the postulation of mental states produced by the action of the brain—carried with it a heavy burden of unintelligibility that has not been successfully discharged by neurophysiology to this day. The great merit of the natural attitude is that it tells its story in the consistently non-causal terms that are appropriate to the kinds of facts it has to relate. By so doing it avoids the strangeness that inevitably infects the dualistic attempt to combine two radically different kinds of accounts with one another. This is not to deny the need to achieve some eventual integration of those accounts; but such a synthesis will not be convincing unless it makes a place for just this obliviousness to the role of the brain that characterizes our normal functioning as human beings.

One tentative conclusion one could draw from this all-too-brief review is that the natural attitude is still in the game as a plausible alternative to both dualism and naturalism. Those theories in their different ways tell us that we are simply mistaken in some of our most fundamental beliefs about ourselves as human beings. Dualism does this by discounting as naive our claim to know the world in perception. But it then generates an even more serious set of puzzles by postulating a mind that is supposed to represent the world, but has no access to anything but its own representations. Naturalism simply denies that there is anything distinctively mental, but leaves unanswered the question of how its own account of the world can dispense with any actual experience of what that world is like.

On the strength of such examples as these, it seems doubtful whether philosophy has any business propounding these peremptory and dismissive judgments on the natural attitude unless it has a manifestly superior set of beliefs to put in the place of those that are being rejected. Even the brief review of dualism in the preceding section of this chapter makes it seem very unlikely that this has been the case historically; and the outlook for naturalism hardly seems more promising. It is certainly not likely, for example, that stronger grounds for affirming the existence of the world will become available from other sources once the testimony of the natural attitude on this point has been rejected. Accordingly, in this study I propose to reverse the traditional procedure and to accept the claim of the natural attitude unless and until it can be shown to be false by proofs that are more persuasive than those that have been offered so far. This means that the natural attitude will not serve simply as a datum for the kind of philosophical interpretation that is quite free to replace it with theses of its own devising. Instead, it will represent in some sense the primordial achievement implicit in having a world at all—an achievement that all philosophical and scientific accounts of our nature as perceiving and thinking beings have somehow to acknowledge and find a place for in the theories they construct.

This may seem an excessively weighty entitlement to confer on what some would doubtless regard as just another theory—a little cruder, perhaps, than its more up-to-date rivals but with no more initial plausibility than they can claim. My reply to this objection is that such a categorization of the natural attitude as a primitive theory has little to recommend it. Theories, after all, have to be tested by their application to a field that they do not simply control so that it can only confirm the theses they propose. That field is what we call "the world" and it can serve this purpose only for beings who may be said to be in it as the world that they have and who share it with others as these others do with them. World, self, and others—these are "facts" that the natural attitude presupposes and that every human being must somehow be familiar with because they are implicit in every form of inquiry and in every human practice. No theory could be confirmed unless the field of inquiry were already ordered by the distinctions they imply. My thesis is, accordingly, that unless a theory reconstructs human being in such a way that this kind of familiarity finds a place within it, it hardly deserves to be treated as even minimally adequate to its ostensible purpose.

This way of putting things will probably not be acceptable to all and certainly not to hard naturalists. Their inclination is typically to hold that whatever emerges from the application of the canon of scientific method to a given subject matter is *ipso facto* the only defensible account of it that can be given and is hardly required to satisfy intuitions from hither and yon that we may bring to such inquiries. What this implies is that in our thinking about ourselves we can theorize as freely and with as little in the way of a prior understanding that guides our efforts as we do in our thinking about, say, subatomic phenomena. That kind of freedom is apparently deeply congenial to many of our contemporaries, as the idea of a prior understanding that cannot simply be junked is not. Against this, I will be arguing that there is an ontological fact—what I will call the fact of givenness or presence—with which we are all familiar and which cannot finally be burked or ignored. As I have already indicated, it is not assimilable to the status of a theory that contemporary thought likes to assign to almost everything it comes across. More than that, these pre-philosophical and pre-scientific ways of understanding perception can be shown to play a role in both philosophical and scientific thinking about human beings. I hope to show, in fact, that such thinking relies on common-sense understandings at various junctures in its argumentation in a way that is hardly consistent with the dismissive treatment that such views so often receive at the hands of the exponents of much more highly theoretical accounts of the same themes.

There is a more general philosophical point that is connected with these considerations; and although it cannot be fully argued here, it is relevant to the kind of justification I would offer for the line of thought I will be developing in this book. We are all familiar with the movement of scientific thought that at least seems to leave behind the common-sense understanding of the world and postulates a domain of smaller and smaller particles to which we have no direct perceptual access. This is the domain in which the natural sciences claim to find the ultimate constituents of the physical world; and it is assumed that, in a "theory of everything," everything would be shown to be constructed out of these elementary entities. In order to do that, physics would have to show that human beings themselves are constructible on these principles. But that means not just human beings as physical organisms, but human beings as inhabiting a world in which there are qualitative features —those of color, for example—that are plainly different from the electromagnetic radiation that, according to physical theory, is supposed to produce them. There is no use dismissing these features as "mere appearances" because even if they are produced in us by certain frequencies of radiation, they are unmistakably real as just such "appearances." The point here is that physical theory has no way of accounting for them, either by showing why this band of radiation produces color at all or why one specific color is produced by a certain wavelength and another by another.

Color, of course, is only one such feature of what has been called—not altogether felicitously—the "manifest image" of the world from which physical theory tries to abstract as completely as it can. It is true that certain phenomena in quantum physics have made it appear that it is not really possible to eliminate "subjectivity" and thus the human subject from the kinds of facts with which physics deals.[4] The point I want to make is a different one. It is that there is good reason to conclude that physical theory is unable to find its way back, under its own power, to its own starting point in the world of the manifest image which is also the locus in which the observations are made that confirm or disconfirm theoretical conjectures. It cannot, in other words, reconstruct in its own conceptual idiom the human subject that is the protagonist of physics as an epistemic enterprise or the kind of world in which that subject lives. But in that case there is evidently an asymmetry in the relation between the common-sense understanding of the world and the scientific one. It consists in the fact that it is possible to get from the former to the latter but not the other way around—not from the scientific image to the manifest one. This means that a "theory of everything" constructed along these lines will not really be a theory of everything.

This fact can be interpreted in a variety of ways, most of which are irrelevant for the purposes of this inquiry. What I think is important is the fact that the kind of interdependence it asserts between the two "images" does not justify belittling either one of them in the interest of the other. They are in a deep sense complementary to one another; and if that fact confronts us with puzzles to which we have no answer, then that is a fact we have to live with instead of trying to plow under one or another of the terms that put us in this predicament.

III

There are two further points that need to be made about the natural attitude. One of these has to do with the surprising resemblance to naturalism that it can seem to have. For example, when naturalism says, as it does, that there are no minds and that there is nothing but the spatio-temporal world which it calls "nature," one might ask how different this is from the thesis of the natural attitude about the existence of the world. Both appear to postulate the reality of a world of things and to maintain silence about anything that would not find a place in such a world. Sometimes, in fact, it can look as though "naturalism" and "the natural attitude" have more in common than just names that are based on the word "nature." It might even be thought that in moving beyond dualism philosophy has come back around to its own starting point and that naturalism is really just a more philosophically sophisticated version of the natural attitude.

What is true in this is that the natural attitude is, in a certain way, ambiguous precisely because it is pre-philosophical. It does not, in other words, emerge armed at all points to do battle with other "theories"; it simply affirms what it has never occurred to it to deny. As a result, it is vulnerable to interpretations of what it is really saying that take its affirmations to new and unfamiliar levels of specificity and implication. This is especially the case in everything that has to do with the natural attitude itself as the affirmation of the existence of the world and with any implications that may flow from the fact of there being such an "attitude." By contrast, naturalism is anything but naive. It emerges from its contest with dualism bristling with arguments and counter-arguments and altogether hyper-aware of all the things that the natural attitude has never thought of. It is, in short, motivated by philosophical considerations as the natural attitude is not. For both, the world is an aggregate of entities; but the silence of the natural attitude about how it comes to be accessible as the world it is, is in a true sense naive, whereas that of naturalism amounts to a principled denial that there is any real question that needs to be asked about this. One thing does need to be noted here and that is the fact that, to the extent that it remains philosophically un-self-aware, the natural attitude has a disconcerting tendency to pass over into a quasi-naturalism that does not seem to understand what differentiates it from naturalism as such.

This is where the second point about the natural attitude comes in. In his *Phenomenology of Spirit*, Hegel postulated a form of consciousness that he called "natural" and that was defined by its confident assertion of the existence of a world of particular things in much the same way as the natural attitude is. What natural consciousness needs to learn is that this truth that it insists on presupposes, among other things, the reality of the self that asserts it. It needs, in short, to become, for itself, the self-consciousness that it already implicitly is. Instead of simply suspending natural consciousness as Husserl does, the *Phenomenology* follows the route one would have to travel if one were to bring all these presuppositions to the level of such an explicit self-understanding. This study does not set itself any goal as ambitious as that. What it does attempt is to show that the natural attitude presupposes something that I will call "presence" as a condition of the possibility of its own relation to the world whose existence it asserts.

The conception of itself that I will try to draw out of the natural attitude is very different from anything that naturalists would find congenial. They typically assume that this whole question of how access to the world is achieved can be dealt with by treating it as though it were an internal feature of the entities in question—that is, once again, human beings. It is as though some event or process inside each of these entities could constitute the world's being there for it. Since any such process must be readily assimilable to the fundamental ontological type of the entities and events that make up the natural world itself, the naturalist finds it hard to discern any deep philosophical puzzle about this. He simply insists that the suppression of mind makes the human being a body and that this body is a physical system. Plainly, there can be no place in such a world for anything like an appearing—a showing-itself to anyone—of the world or of anything in the world. It follows that, as fully incorporated into nature, our ontological status will once again be that of things. The main difference between naturalism and dualism is thus what *kind* of a thing we are—physical or mental. It doesn't much matter which kind it is, though, because in the one case its awareness is confined to itself and in the other there is no awareness at all. Neither kind of thing can *have* a world as something that is there *for* it.

The issue about naturalism that is to be resolved in the chapters that follow is thus significantly different from the one that usually dominates discussions like this one. Usually, the challenge that is made to naturalism is to show that certain qualities or experiences that do not readily fit into the physicalistic concept of human nature—pain, for example—cannot really be accommodated there. What neither the naturalist (nor, for the most part, his critic) seems to be puzzled by or even to notice is the fact that things in the world, within which everything is declared to be physical in character, are *there* in the sense of being disclosed. One principal obstacle in the way of such an acknowledgment is the fact that the world itself is supposed to be the creature of theory and, as such, not the kind of thing one would think of oneself as just coming upon. As one scientist has put it, there is "subjective experience" and "all else is speculation"—in other words, something we have spun out of our own heads.[5] What I want to show is that naturalism fails precisely because it cannot find a place within its categories for this fact—the "thereness" and, more generally, the open character of the things that are within the world.

It will be apparent from what has been said in this chapter that the focus of this inquiry into naturalism is no longer the status of mind; it is that of human beings conceived as unitary entities in the sense of not being a compound of mind and body. As a consequence, the central issue it must resolve has to do with the way a human being is in its world with other entities, and whether it interacts with its environment in ways that are susceptible of being explained, as naturalism requires, entirely in causal terms. Since causality is an external relation and not a nexus of meaning, it requires that a human being as a term in such relations be conceivable independently of the other terms with which it is associated. These are the entities that act upon it and may in turn be affected by its reactions. My argument will be that the way human beings are in the world with other entities cannot be understood on the model of physical systems. We *have* a world in a far more intimate way than can be rendered in terms of a relation like that of causality. But this means that the naturalistic project miscarries, not because it ignores some transcendent metaphysical truth, but because it grossly misrepresents the character of our way of being in the world with other entities. I will also suggest that it is able to pass off this account as accurate only because our familiarity with our own way of being in the world is as inarticulate as it is profound.

3

THE REJECTION OF THE GIVEN AND THE ECLIPSE OF PRESENCE

I

In the preceding chapter, it was made clear that the focus of this inquiry would be on the way human beings are in the world with other entities. As a preliminary indication of what is distinctive about this way of being in the world, the concept of the given was introduced and with it, as a kind of synonym, the idea of presence. In this chapter, the most conspicuous and arguably the most fundamental modality of givenness/presence will be taken up: perception. Perception is not the only modality that presence can assume; the next chapter will give an account of another that is fundamental to language. It is, however, the most obvious example of the way human being transcends the body as a physical system. In perception, as we ordinarily understand it, something that is typically outside our bodies (and outside our "minds" as well) is supposed to be directly given or present to us and this presence is what is achieved in perception. It is also deeply familiar to us and is normally taken to be wholly unproblematic. Nevertheless, as has already been explained, in almost all scientific and philosophical accounts of perception, taking it at face value has been judged to be unacceptably "naive" by reason of the unhesitatingly realistic character of such a view.

Sometimes the remedy proposed has been to replace the object in the world with some pattern of "experiences" and thus to bring the whole transaction indoors by treating it as an intramental event. When this is done, perception is construed as being internal to the perceiver and to his mind and not as his way of being in the world. This is the way of dualism as discussed in Chapter 2. Another—typically unavowed—way of dealing with the realism that characterizes our normal understanding of perception is akin to what Whitehead called "the fallacy of misplaced concreteness." This is a fallacy that is committed when the prominence and concreteness of some element in a situation misleads us into assigning a more fundamental importance to it than is really warranted. In the case of perception, it is the role of our sense organs as parts of our bodies that becomes the principal focus of interest and distracts attention from their function as necessary conditions for our having a world—admittedly in a way that we understand only very imperfectly. Unfortunately, because having a world is so familiar to us, we do not really notice it and certainly do not "thematize" it. Instead, by concluding that whatever occurs in perception must be something that occurs in us—some process taking place in our sense organs and our brains—we really make this achievement invisible to ourselves. In this way, attention is diverted from having a world to something that seems to lend itself more readily to explanatory procedures with which we are already familiar. At the same time, as has been pointed out, we implicitly equate the fact of presence with the existence of the objects we perceive and thereby insulate ourselves from any sense of wonder at the fact that they are there *for* us. We need, surely, to be emphatically reminded of how extraordinary it is that these solid-looking beings—we ourselves and others like us—open on the world that surrounds them. It is, one might say, as though all that opaque flesh that might have been expected to block any access to the environing world had been rendered transparent and not just transparent but invisible.[1]

Finally, there is yet another reason for giving perception pride of place in such a study as this. It is much neglected in contemporary philosophy; and the reason for this neglect may well be that it constitutes the most substantial obstacle in the way of one of the principal tendencies in contemporary thought: the recrudescence of something very like rationalism. Rationalism means, as it always has meant, the priority of conceptual thought over any actual experience of the things it deals with. At present, the "reason" of classical rationalism takes the form of "language" and "theory"; and the priority of these elements in cognition is often associated with the thesis that all experiential data are somehow already in the service of a theory. There is, of course, this difference between the classical and the current versions of rationalism, namely, that the latter permits a much greater degree of freedom to design the world as one pleases than the former ever did. The priority assigned to language in contemporary philosophy lends itself admirably, it seems, to many very different strategies of argument; and it is capable of turning even apparently unassimilable data into facts that are at least compatible with almost any purposes. Even so, because perception at least seems to confront us with a world that theory did not create, it offers a resistance to the sovereignty of theory that is not really welcome.

II

The tendencies just described play a considerable role in the way perception has traditionally been conceived in philosophy and psychology. In this section, that way of conceiving it will be reviewed and special attention will be given to the linkage of the concept of perception with that of sensation. Sensation as a modification of our mental state by a process originating outside the mind is supposed to be the qualitative core of a perception. It is, accordingly, what is in a proper sense "given" in perception. By contrast, the extramental object of a perception would, on this view, be posited in an inference or judgment and is, therefore, not itself given or present in the way a sensation is supposed to be. This givenness is what finds expression in the term "sense-datum" that has recently been favored over "sensation" by philosophers; and it will be the term used in the analysis that follows.

It has already been made clear that naturalism is determined to make no use of this concept and, if possible, to strike it from the list of concepts that have any place in a science-oriented philosophy. Unfortunately, this zeal to dispense with sense data has meant that there has not been very much attention to the concept of the given as such, as distinct from the various logical vices that are attributed to the sense-datum as an object of knowledge. One thing does seem clear about the notion of the given, however, and that is that it is designed to express the fact of something's appearing to someone, whether as something visible or audible or in some other sense modality. When this is borne in mind, it becomes evident that to reject the idea of anything's being given must have some rather serious consequences. How, for example, could we ever come to be familiar with such a distinction as the one between red and green if nothing red or green ever puts in an appearance? More generally, if nothing is ever given in the sense of being present to someone, how could there be anything like observation in which, surely, what is observed must somehow be present to the observer? The philosophical motives that could justify making perception and observation walk the plank would have to be weighty indeed.

Those motives have been mainly epistemological. The special quality attributed to the given has been the certainty of the knowledge we have of it. This certainty is supposed to be linked to the "immediacy"—the non-inferential character—of the given; and it is the key to the philosophical motivation behind the use of the notion of the given when it was more popular than it has since become. That motive was to provide a secure epistemic foundation for empirical knowledge generally. What was given was conceived as a sense-datum and this, in turn, was understood as what is immediately present to the mind in any form of perceptual consciousness. Its function was supposed to be to serve as the object of primary empirical apprehensions. The assumption was that these could not be mistaken because their object —sense data—could not *be* otherwise than they appeared. As it turned out, however, logical peculiarities that derive from just this vaunted incorrigibility have made it seem quite unsuited to the function that was thus assigned to it.

The paradigm instance of a sense-datum has always been something like a red color patch; and philosophers have rightly emphasized the difference between such a pure experience of quality and a color property. The latter is an attribute of a *thing* of some kind; and, like that thing, it can look different under different conditions—in different lighting, for example—while remaining the same property. By contrast, a sense-datum altogether lacks the character of a perduring property and is not even susceptible of being re-identified as the *same* sense-datum. What all this has been taken to imply is that the given, conceived as a sense-datum, lacks the kind of integrity and continuity that would enable it to serve as the foundation of empirical knowledge. But, as so often happens, the baby has been thrown out with the bathwater—the concept of the given with that of the sense-datum. The former is the idea of something that is given or present to another entity; but it is not necessary that what is so given be as impoverished in its ontological constitution as a sense-datum is.

Traditional ideas about what a sense-datum must be like are traceable to psychological assumptions about how a stimulus could be registered, first, by the organ that is sensitive to it and then by the consciousness in which it is to appear. The atomic character that the resulting sense data were supposed to have, made it appear that what is so given would be identical with its appearance so that a mistake about it would be impossible. This way of conceiving it was supposed to guarantee that it would be free of any tincture of interpretation and be the pure imprint on the mind of something acting upon it from without. Against this, it has been argued that there is always some context in (or background against) which we experience something and that any quality we perceive varies subtly with that context. To put this in a somewhat different way, although some philosophers have thought that our familiarity with our own allegedly internal states does not *ipso facto* commit us to anything as unwieldy and unmanageable as a world, those states have proved to be very like the proverbial camel's nose. Once it is inside the tent, the rest of the camel follows. So it is, too, with "sense data" which it may be more appropriate to conceive as fragments broken off from the great cosmic process than as fully constituted entities in their own right. This is because their pristine character invariably turns out to be compromised by some affiliation with things and events that have an unmistakably worldly character. The moral to which all this points is that there is no way of keeping the world at arm's length. Although we may wish to deal with the "external world" only through its diplomatic representatives, the reality of the matter is plainly a face-to-face encounter.

If the sense-datum is already, as one might say, a piece of the world, then our way of having a world must be, quite literally, *in medias res* without any logically prior experience of "pure quality." In these circumstances, a natural inference might have been that it is things—material objects—that are given and not sense data. The conclusion most philosophers seem to have drawn is very different. It is that if the given cannot be a sense-datum, nothing at all can be said to be given and the concept of the given itself should be abandoned.[2]

In the background of this second inference, there is the idea that, in order to give an account of mental fact, one must postulate a special set of objects.

THE REJECTION OF THE GIVEN AND THE ECLIPSE OF PRESENCE

These objects would compose a kind of mosaic and this mosaic would constitute our conscious experience and thus pretty much what we mean by "mind." The difficulty about this has been to explain what it is about these objects that makes them phenomenal objects—objects that appear. In the case of the objects we encounter in ordinary experience, the fact of appearing seems quite distinct from any property we can assign to them; and so it does not seem that they can themselves account for their own phenomenality. This is probably why these objects have usually been conceived on the model of sense data; but there is good reason to doubt whether the latter are any better qualified to answer this question than full-blown spatio-temporal objects are. It is true that strenuous efforts have been made to make it appear as though sense data are tied so closely to the experience in which they occur that they cannot do anything except appear. For one thing, they famously do not have backsides and consequently have to be identified with the "side" that faces the being to whom they are given or present. What this does, however, is simply to produce an object that is so mutilated one can hardly think of it as a sensible—something that could have a career on its own, apart from any experience in which it might figure. Worse still, this mutilated object does not enable us to explain the fact of appearance any better than a normal object does.

It is surprising, in retrospect, that this rather bizarre conception should have had as long a run as it did. Even so, the day of reckoning eventually arrived. It came when the very strangeness of sense data that was supposed to enable them to function as the grounds of their own phenomenal status led to their being shown the door. It is a remarkable fact that, in the virtuous enthusiasm that accompanied their departure, no one apparently noticed the gap it created or felt the need to make some other provision for dealing with the fact that something has to be given. It was as though the philosophers who dismissed them were so heavily committed to the idea that the fact of givenness can be accounted for only by postulating some specially designed set of objects, that the fact of givenness itself as something's being there *for* someone was altogether missed. As a result, when sense data as phenomenal objects had to be given up as a way of accounting for psychological or mental fact, only physical objects remained. If anyone wanted to say that we see or otherwise experience something, that something could only be a physical object.

Conceivably, that might have been a positive development, if it had been noticed that a reversion to the natural attitude was effectively taking place. That would have required a willingness to allow a material object to be present, and that was not forthcoming. Instead, a conception of perception emerged that simply dispensed with anything like the givenness or presence of the object that we say is perceived. Perception became something that takes place in the brain of the perceiver—never mind how or where, exactly. For its part, the nominally perceived object remained in its place—usually outside the body—although how our knowledge of this fact could be accounted for remained a mystery. There are, to be sure, physical and neural transactions between the object in question itself and its neural representation, but on this view the object does not and need not put in an appearance. The idea that somehow the object has to be present to the perceiver if anything that deserves to be called a perception is to occur is simply dismissed on the grounds that appeals to what we imagine ourselves to know about perception before neuroscience takes over have to be disallowed. They are just so much folk psychology and, as such, have no standing in a proper scientific inquiry.

The reasoning here seems to be that, as in Locke's example, what we see cannot be an object like Worcester Cathedral, since it could hardly fit into either a mind or a brain so something else that can has to take its place. The point that needs to be made in rebuttal is that there is no question of anything's being *in* either the mind or the brain, whether it be the object in question itself or some representation thereof. Instead, the issue is whether something—a cathedral or a mouse or a flower—can be given or present *to* another entity in a sense that does not entail its being in it. That is what the natural attitude accepts as being the case; and if that, too, is to be declared impossible because there is no physical explanation for any such fact, then we will be back in the paradoxes of either representational dualism or a naturalism that denies that anything needs to be given or present.

The connection between the rejection of the given and hard naturalism now begins to emerge. In the dualistic picture, what was given—the sense-datum or *quale*—was the only phenomenon, the only entity that actually showed itself, albeit intramentally. Linked to this idea was the thesis that the perception of objects outside the mind is mainly a matter of judgment. In judgment, a concept has to be applied to what does appear, that is, to the sense-datum; and this makes perception more a matter of inference to the existence of an "external" object than a direct apprehension of it. This is an inference for which what is given serves as a kind of premise. But when the given was rejected, perception lost the one element of qualitative presence it had. The result was that it became entirely a matter of inference and belief—a discursive operation in which concepts and the language that is their vehicle have the principal roles. When it was further determined that language use itself is a physical transaction, the case for naturalism was complete.

Since naturalism cannot do without the observation of things and events in the world, the question must be whether observation and perception, generally, can really do without something that is given. It does not seem to be a very hazardous inference to claim that if one sees something, something must be in view—in one's field of vision—and, as such, present to a perceiver or observer. Nevertheless, an acknowledgment of this fact has been blocked by another line of argument whose focus is on the linguistic rendering of what is supposed to be there before us. It is pointed out that when we observe something, we approach it with questions about it that we ourselves design

THE REJECTION OF THE GIVEN AND THE ECLIPSE OF PRESENCE

and these dictate the description we give of the things in question. In this sense, it could be said that we bring something of our own—a kind of hypothesis—to the observing we do; and that has seemed to many philosophers enough to call into question the idea of a simple presence of things to us. It is as though by setting the context in which information is acquired, our questions and the descriptive apparatus that informs them somehow pull the things we see into the orbit of our lives and thereby make their "presence" more than a little equivocal.

It has been suggested by Heidegger that this is what is implicit in the concept of perception itself, a word that has a Latin root in *capio* which means "take." But does taking, however construed, necessarily cancel out the fact of presence altogether? Things, after all, do not identify themselves nor do they propose questions of their own that we might properly ask of them without violating that identity. We are left to our own devices when it comes to designing such questions. Where, one may ask, do the answers we get to these questions come from if not from the world we are in? Again, even if our interest in the world and our mode of interrogating it are typically appropriative, are not the properties that make a thing suitable for the use we make of it real properties of that thing? If they are, it would not be as though our becoming familiar with things around us in some context that reflects our interest in them fatally compromised their independence from us or indeed from the names we give them. If they work for us in certain ways, that must be part of their nature and an index of what they are; and as such they are certainly not our creations. This is surely what is expressed by describing them as being present to us. If the suppression of the concept of the given were to deny us the ability to distinguish between what we make and what we find, it is hard to see how that could constitute a philosophical advance.

There is a wider question here that is suggested by the way naturalism deals with the concept of the given. It concerns the whole relationship of naturalism to epistemological issues and especially to those having to do with evidence. In its critique of dualism and mentalistic concepts generally, naturalism appeals, in the name of science, to the idea of evidence and argues that there is no real evidence for the reality of the kinds of states that are called "mental." But evidence has to do with occasions on which the subject matter of an inquiry presents itself (or fails to do so) to those engaged in that inquiry and thereby tends to confirm or disconfirm some claim that has been made about it. When the requirement of this kind of publicity is applied to the study of the mind as conceived by dualism, a difficulty immediately becomes apparent. Because there are many minds and all mental states are the mental states of some individual mind, the relation of any given mind to other minds and their states must be problematic for anyone who operates on naturalistic principles. Dualism conceives minds as being inherently private and their contents as being open to inspection only reflexively—that is, by an individual mind's turning its attention away from things and events in the "external" world and toward its own states and acts. The issue this raises is why anyone else should give credence to reports of the mental states of others since these states can be experienced by only one person. It has sometimes been thought that the only remedy for this uncertainty would be to require that it be possible literally to look into another person's mind. Plainly, though, it would be impossible to comply with this requirement and probably even to understand what it calls for. In the eyes of naturalism, however, the fact that this cannot be done constitutes a crippling limitation to which the ideal of publicity is subject in its application to mental fact.

Altogether, it would seem that the idea of evidence is very closely associated with that of experience and thus with the very notions to which naturalism is so opposed on the grounds of their mentalistic character. An appeal to evidence must, therefore, seem rather strange when it comes from those whose ontology does not make a place for anything like the givenness that is at the core of the common-sense view of experience. There is also the fact that, if this line of thought is pressed far enough, each of us would be left with the sense that the only perceptions that count are his own; and in that case the distinction between something's being public rather than private would have to be called into question. That distinction, after all, presupposes that the perceptions of many people have been accepted and have been in sufficient agreement with one another to issue in something that can be proposed to still others as "objective" fact. The trouble about this is that unless some meaning can be given to the notion of what is public by acknowledging the equivalence in principle of observations made by others with one's own, the public and thus privileged character of scientific fact itself will be left entirely unexplained.

It thus appears that the difficulty about the given here is not in the first instance epistemological as it was in earlier critiques of dualism. It is ontological in the sense that, because perceptions have been supposed to be experiences and as such constitutionally private and inaccessible to others, they have been ruled out in advance by the naturalistic assumption that scientific fact must be objective physical fact and, more generally, that what is public must be physical. But if naturalism has committed itself to the public world of science and what is public necessarily involves the idea of many observers who accept one another's observations, it does not seem that it can simply abandon the idea that there are many perceivers and not just the self. In that case, however, naturalism cannot simply use the dualistic concept of perception for its own polemical purposes as it has traditionally done. It has to come up with another concept of perception of its own; and it is clear that it will have to be one that makes a place for presence.

III

Against the background provided by the preceding account of the rejection of sense data, it should now be possible to understand the way hard naturalism deals with perception. In keeping with its assumption that dualism is its only serious rival, it insists that to perceive cannot be to be in a mental state. It must, therefore, be something that the body does or something that occurs in it. "Body" here is to be understood as denoting a physical object of a certain kind with the properties that are recognized by the natural sciences. The relations in which this object stands to other objects are spatio-temporal and causal. Perception has to be understood as an event caused by a physical stimulus that impinges on the sense organs of the human body and thereby initiates a process in the central nervous system and the brain that eventually issues in overt bodily behavior. This whole process can be described entirely in physical terms and this means in the language of physics and chemistry and any other natural science that uses the concepts of those foundational sciences to study human beings.

The import of this thesis can be clarified by an example. Consider the case of poker chips. Suppose that we have a pile of them and that they are of different colors. We might ask someone—let us call him X—to sort them out by color, placing all those of one color in one stack and so on. Everyone would agree that if X completes this task successfully, that would be good *prima facie* evidence that he has perceived these chips and that he can discriminate their colors. The naturalist wants to say something more, namely, that X's behavioral performance in sorting out these chips is the kind of thing that his perceiving them is quite generally to be understood as being. Normally, we would probably say that seeing the chips is a necessary condition for sorting them out and that a distinction can be made between perceiving such objects and anything that we may then go on to do with them or about them. But, for the naturalist, perceiving has to be something that occurs in our bodies or that our bodies do, so this distinction has to be challenged. Above all, one must not allow perception to be an internal episode in the bad sense—something happening in the mind.

There are countless objections to this kind of analysis. Among these, the most general might be simply to point out that this account offers no way of picking out what we ordinarily think of as perceptions from the other reactions to the countless causal influences from without that are constantly impinging on our bodies and producing their effects there without any awareness on our part. Indeed, the whole idea of making such an intimate connection between perception and some kind of behavioral response seems highly questionable. Most perceptions are not accompanied or followed by any action at all on the part of the perceiver. The naturalist's answer to this is that something is always happening in the brain and that, even though no overt action is performed, beliefs are being formed as a result of these brain-events. This leads to the claim that perception is the formation of beliefs about whatever we perceive and that perception itself is belief as a dispositional state that at some time or other manifests itself in a variety of behavioral ways.

Since belief is arguably a mentalistic concept, this thesis may seem to go in a different direction from the naturalistic line of thought. In the naturalistic idiom, however, an appeal to belief serves only the purpose of linking perception with behavior. Thus, if I see that my car is parked too close to the one next to it, I may be said to have learned something and what I have learned will manifest itself in the way I back out of my parking space when I return at the end of the day. Naturalism does not hypostatize this belief as some sort of intervening mental state; it is understood simply as the wider coordination of what I do with what I have been able to learn through perception. As such, it can be identified with some modification of my brain state that occurred when I saw how my car was parked.

This thesis about belief dovetails neatly with the assumption that perception is something that occurs in the human being that perceives. The concept of belief was not originally formed for the purpose to which it is put by naturalism; but certain of its logical features have a curious affinity with the naturalistic program. A belief, for one thing, more or less openly declares itself to be something that could prove to be false and it may, accordingly, say more about the state of the person whose belief it is than it does about anything else. It is also implicit in the terminology of belief that, at least for the time being, what we believe is not confronted by what it is about in such an unmistakable manner that there will no longer be any room for conjecture about it. But if we distance the object of perception from ourselves in the way we do the objects of belief, we will have to accept a number of rather peculiar consequences. I will have to say, for example, that I *believe* that I am now in my own house even though there is nothing to suggest that I am anywhere else and everything around me is what I have seen and would expect to see in my own house. This makes it sound as though this "belief" of mine were on a par with my belief that O.J. Simpson was guilty. The point is not that I could not be mistaken in the one case and can in the other; it is rather that to speak of belief in the absence of any plausible grounds for uncertainty results in a false assimilation of such cases to those in which there simply is no occasion for uncertainty or doubt.

Among the behavioral manifestations of our beliefs are utterances in which something that was learned from an original perception may be expressed. Thus, in the course of his sorting out the chips, our man X might say "This is a green chip" while holding in his hand a green chip; but something of this kind could also come to expression long after the occasion on which the relevant belief was formed. The idea that such an utterance would constitute physical behavior in the required sense will be examined in the next chapter. Suffice it to say, for now, simply that all these behavioral sequels might equally well *not* ensue and so their adequacy as renderings of what perception consists in must remain moot. Perception, by contrast, is not conditional in this sense, but categorical. At any given time I either do or do not perceive a given object and so whether or not I perceive something cannot be made conditional upon the occurrence of some eventual behavioral performance. The latter may, of course, serve as the basis for someone else's inference that I have in fact perceived something; but that is a different matter.

The implausibility of a behavioral account of perception can be brought out in other ways as well. The decisive consideration in its favor appears to be the fact that such behavioral sequels are all that other people who observe this perceiver can themselves perceive. But when we say that these others observe the overt manifestations of beliefs acquired by our subject in the course of perceiving an object A, we are surely not treating their perceptions in the same way as we treat his. We are not talking about their behavior, verbal or otherwise. Instead, we are talking about what they are able to perceive, namely, X's body and its movements as well as the fact that he has correctly separated the chips from one another by color. By contrast, X's perceptions *are* being equated with his behavior and if something creeps into the description of his behavior that goes beyond the movements of his body, that will be the responsibility of the observer who is perceiving X as he goes through this drill. This raises the question of why we should be unwilling to say that when X examines these poker chips, he is not simply producing behavior that in fact accurately separates them by color. It seems entirely natural to say that he sees the chips in the sense of having them in view and that this is a condition of his being able to do the sorting correctly, whether he goes on to sort them or not. But if we decline to acknowledge this fact, we will be compelled to treat the perceptions of the observers in the same way. In that case, there will be no way their perceptions can be perceptions of what X has accomplished or any way of determining whether he has done it correctly.

By way of a remedy, it would not do any good to summon still other observers to observe the first set of observers since the same fate would befall them and an infinite series would threaten. And if the naturalist were to reply that he is not talking about other people at all and that everything about them would have to be dealt with in just the same way as the perceptions of X, we would have to ask him where he is speaking from and whether he himself is not the relevant observer for these purposes. Surely, he has seen and is now imagining seeing someone who is engaged in tasks like that of sorting poker chips. Is he willing to say that his own perceptions of such a person are to be identified with his own behavioral responses to it? This is hardly likely; but, whether it is acknowledged or not, it is clear that once again any observer of this scene who can testify to what X has accomplished will do so out of the natural attitude. What typically happens is that the non-behavioral content of his perceptions is tacitly read into those of the experimental subject. Indeed, it is only in this way that the extreme implausibility of the hard naturalistic thesis about perception can be got around.

Sometimes attempts are made to avoid mentalistic construals of verbs of perception by claiming that "see," for example, is an achievement word and that it functions as a signal that something has been found. Even though signals often do form part of behavioral routines, they do so in a way that is not helpful for these purposes. They have, after all, to be perceived by both the person who gives the signal and the one who receives it. Unless one could show that these seeings as well are susceptible of being dealt with in terms of the behavioral paradigm, it doesn't seem that the latter's scope would be large enough to prove anything much. It must also occur to us that in order to find something we have often to look in many places. "Look" may have a more behavioral feel to it than "see"; but is it really possible to talk about looking for something without implying that seeing is involved—seeing things that are mostly not the one we are looking for? But in that case the achievement aspect of seeing begins to seem incidental and does not move visual perception any closer to the behavioral paradigm.

There is another related point that weighs heavily against such a treatment of perception. When we do react to something that we perceive, it is by virtue of some feature of or fact about the perceived object. It may be valuable or dangerous or sacred and in each case an appropriate response will be forthcoming. None of these responses, however, can duplicate the richness of the perceptual object which always has characteristics that are irrelevant or indifferent to the interest we take in it. This fact does not keep them from figuring in our perception of this object. In this respect, our responses to a perception are abstractive in the sense that they pick out one or more properties as the only ones that really count. It may be noted that actions are in this respect like the statements we make about what we see or otherwise perceive. They may succeed in expressing what the perceived object *is* in terms of some system of classification that is relevant in a given context. At the same time, however, they will unavoidably also omit a great deal that is available to us in perception. Because they do so, there can be no question of identifying any action motivated by a perception with the perception itself or with any disposition to perform that action or other actions. It looks very much as though this whole idea is a misguided effort to make the sound thesis—that things are pervasively perceived as *pragmata* and *manipulanda*—work overtime as a definition of perception itself.

A fall-back position has already been sketched to which naturalism can withdraw in the face of these objections. Instead of tying perception to overt behavior, it can argue that perception is an event in the central nervous system that may very well not issue in any overt observable behavior at all, but that would, nevertheless, make such behavior possible if appropriate circumstances were to arise. Although the main discussion of the role assigned to the brain in hard naturalism will be reserved for Chapter 5, some observations on this line of thought as it relates to perception can be offered here. In particular, the unique obstacles that perception places in the way of hard naturalism need to be understood.

Perception is spoken of as an event that takes place in the brain; and naturalists do not hesitate to claim that certain events occurring in the brain

THE REJECTION OF THE GIVEN AND THE ECLIPSE OF PRESENCE

simply *are* perceptions. What is most puzzling about this is how people who talk about perception in this way can reconcile what they say with the fact that what we perceive typically lies outside our bodies and our brains. How, then, can a neural event that occurs inside our skulls constitute a perception of something that is so clearly separate from it and has such completely different properties? There is, of course, a connection in the line of physical causation between the external object that we say we perceive and the brain state that comes into being when our sense organs are stimulated by light reflected from that object. That brain state in turn is connected to certain movements of our limbs that arguably respond to something about this perceived object. But if, as has already been shown, that behavior understood simply as a series of movements does not constitute a perception of that object, then how can the electrical and chemical events in the brain that precede this behavior be said to be a perception of that object? Or, again, how can perception as a series of physical events in the brain of a perceiver somehow go back up this chain of causation and bring it about that an object outside my body becomes visually or otherwise present to me? If these questions cannot be answered, then it must follow that no perception that attains its object ever really occurs—a hard pill to swallow, surely.

It is possible that this puzzle has been generated by the way hard naturalism has taken over some of the conceptual apparatus of dualism. If so, it seems very likely that certain features of the concept of consciousness are at the center of the puzzle. As its etymology indicates, "con-scious-ness" is the concept of "knowing with" and it is no accident that its alternate form, "conscience," expresses our awareness of the moral quality of our own thoughts and actions—that is, of what is going on in our minds. In its original mentalistic/dualistic setting, consciousness—"the light of the soul"—was thus at one remove from the primary representations in the mind of things in the world and at two removes from those things themselves. In these circumstances, it is evident that the concept of consciousness might more appropriately be understood as a concept of self-consciousness since it is an awareness of what is happening in the mind and only very indirectly of what is the case in the outside world.

What is even more interesting is the fact that hard naturalists have not simply jettisoned this concept as they might have been expected to do. Instead, they describe themselves as *looking* for it and as looking for it, of all places, *in the brain* which for them has replaced the mind. This search constitutes a bit of a puzzle on its own since, in the absence of a discovery of something in the brain, one wonders what it is that the naturalist can be talking about when he invokes this term even before he has discovered anything in the brain that can plausibly be baptized "consciousness." Is he admitting that he already has some understanding of what consciousness is independently of what brain physiology is expected eventually to tell him? What could be the source of this understanding which he evidently shares with a great many other people who typically know nothing at all about the brain? And, most important of all, what is it that the brain can be shown by neuroscience to do that justifies our assigning functions like perception to it, with which we already have some familiarity?

It thus suggests itself that the peculiarities of the original concept of consciousness that have just been described have carried over into the naturalistic way of conceiving perception. If so, that would indeed tend to make perception a distinct event that takes place in the "mind–brain." But this explanation will work only if one bears in mind that, like everyone else, neuroscientists and their philosophical colleagues live in the natural attitude. When they perceive something, they are therefore perfectly sure that there is an object of a certain kind in their vicinity. At the same time, they study the brain and isolate events there that are evidently necessary conditions for such a perception as this. The fact that the role played in the process they study by the object that is perceived precedes the perception itself does not call into question the assumption that that object is what is perceived because they already know this on other grounds. The fact that those other grounds cannot be fitted into their conception of that neural and behavioral process is either not grasped at all or is finessed by an unavowed shift to a dualistic and representational picture in which what is present to consciousness is not identical with the object outside the body. At the same time, attention is diverted from the flagrantly non-naturalistic character of any such conception by talking instead about what the perceiver *says* about the perceived object. This is conceived to be the naturalistically legitimate continuation of the physical process in our bodies that has already been recognized as bringing about the perception to begin with.

The moral of the tale thus appears to be that it is the fate of perception, when it falls into the hands of scientists or philosophers, to wind up encapsulated within an enclosure of some kind, whether it be a mind or a brain. Different as the one is from the other, the effect of the encapsulation is the same in the two cases. It is to sever the relation between perception and its object. What I am suggesting is that this conceptual mutilation carries over from dualism to naturalism together with other concepts like that of representation. These continue to function in much the same way as they did in dualism although with the shift from a mentalistic to a physicalistic ontology their import has been drastically modified. My thesis is that if this kind of continuity between such fundamentally different theories is possible, then the features of the concept of consciousness to which I have drawn attention can carry over as well.

Confirmation for the line of thought just sketched is offered by the strategy that some naturalists adopt when they are compelled to recognize the force of arguments demonstrating the distinctness of perception from any brain process. What they do is to conceive in the narrowest terms possible the consciousness for which they are compelled to make a place. Sometimes this means in terms of "what it is like" to be a human being or a bat or whatever.[3] This inevitably suggests that a state of feeling of some kind is what makes perception something more than a brain-event; and if a state of feeling is acknowledged, there must be a consciousness to which it belongs. Indeed, many philosophers appear to have a settled preference for treating feelings— usually, feelings of pain—as the paradigm case of consciousness. Of course, if it is conceded that feelings are not reducible to physical states, the main ontological thesis of naturalism has to be set aside. Even so, there is a sense in which this concession can at least appear to maintain the superficially similar thesis that consciousness is *in* a human being even though not in the brain in any literal sense. In this way the character of consciousness as presence does not emerge as starkly as it would if its transcendence toward the world were openly acknowledged. In effect, mental fact is being interpreted in terms of consciousness and this means as the self-consciousness that was implicit in that concept since its inception. Altogether, there is good reason to think that for these and other reasons the concept of consciousness is quite unsuitable to service as the master concept for the characterization of human being.

Implicit in much of what has been discussed up to this point is a major thesis of the naturalistic account of perception that needs to be noticed and appraised here. It is the claim that perception has a causal character and that we are caused to perceive what we perceive by what we perceive. This thesis has a certain plausibility since there can be no doubt that all kinds of causal relations *are* necessary conditions for perception and that the perceived object is itself involved in these causal transactions. These mostly occur in the central nervous system and in the brain; and if perception were simply an event that takes place in the brain, a causal theory of perception would pretty clearly be on the right track. It has been shown, however, that perception cannot properly be conceived in a way that separates it from the perceived object in this radical fashion.

The crux of the matter is the fact that a causal relation involves at least two terms; and so the question must be how naturalism can come by these terms, consistently with its own principles. It could, of course, do so by a reversion to a dualistic position since for dualism what we see really is numerically distinct from the object outside our bodies. That picture certainly provides the two terms that are required for a causal relation; but it goes with a set of ideas about the mind that naturalism has repudiated. The issue for naturalism must therefore be to show that the causal character of perception can be asserted without reverting to dualism. Clearly, however, it cannot do this if it persists in a purely physiological-cum-behavioral account of perception that is the only one it can consistently maintain. The causal character of these physiological processes may not be in doubt, but the question will be whether any account that confines itself to them would still be a theory of *perception*. The big fact here is that naturalism is precluded by its own principles from making the required distinction between cause and effect in such a way that the effect is really a perception.

It looks very much as though in making this claim about the causal character of perception, naturalism is really saying no more than that the object one perceives as well as the physical processes that originate in it are necessary conditions for my perceiving it. This means that there will be a logical gap between the physiological story it tells and the fact of the object's being there in perception—the fact on which the natural attitude is founded. Because both dualism and naturalism bypass the natural attitude, they refer to the objects we encounter in two different ways. They are physical objects acting upon our sense organs prior to any awareness we may have of them and they are also supposed to be the states produced in us by the action of those "external" objects, whether as sensations or brain states. What is remarkable in all this is that none of these descriptions that naturalism and dualism give of the effects that are produced in us fit the objects we see or otherwise perceive. It is evident that the objects with which we are most conversant do not conform to the description physics gives of what the world really consists of. It is equally clear that they cannot be appropriately described in the language of sensations as though that were the guise in which they are made to appear to us through the action on us of the entities of physics.

IV

The upshot of the preceding critique of hard naturalism as a theory of perception has been the clear implication that the concept of givenness/presence is still required for a philosophical understanding of perception. It does, however, need to be understood in a way that does not place it uniquely in the service of an intramental conception of perception. In this section I hope to provide a somewhat larger context for such a conception of presence and thereby to lay the foundation for the even fuller account of it as a constitutive feature of human being that will be presented in Chapter 6. In doing so, I propose to use the word "presence" instead of "the given" or "givenness." The primary associations of the latter terms are epistemological and I want "presence" to express something ontological. What that is, will emerge progressively in the course of this section and in Chapter 6; but a few preliminary indications may be in order here.

The most readily available way of coming at "presence" in this altered understanding of it is to say that presence is the condition that accrues to an object when it is perceived. This way of putting things creates a strong temptation to assume that presence must have a relational character and also that perception somehow confers presence upon what is perceived. In some sense, presence undoubtedly does have a relational character; but it would be unwise simply to take it for granted that this "relation" really satisfies the criteria we normally set for something's counting as a relation. Whether it does or not is a matter that will be examined more closely in Chapter 6. The relevant point now is that when someone perceives something, the perceived object is *there* for the perceiver in a way that is utterly familiar to everyone but for which we have very few appropriate locutions. One can say that when that something is perceived, it is "in view" or, more freely, that it "shows itself"; but the expressions that are most readily available—its being "visible," say, or "audible"—simply equate presence with its linkage to psychologically conceived acts of perception.

We have already seen how, by this route, "presence" comes to be conceived in dualistic terms and is acknowledged only as the internal illumination in which the contents of the mind are supposed to be bathed. The cycle of critiques has also been described in the course of which the re-presentations that are supposed to be the interior furnishings of the mind lose their home there and become brain states while the nominally perceived object is reduced to one term in a causal relation that in no way involves anything like presence. In this picture, the only relations in which human beings stand to things in their world are spatio-temporal and causal and this conception of human beings as objects among objects is the one that naturalism makes its own.

The very different line of thought with which the concept of presence is affiliated stems from the fact that as human beings we have to deal with other entities as having a certain character or, differently put, as being such-and-such—apples or oranges, automobiles or earthworms. Sometimes this fact about us is expressed by saying that we deal with things under descriptions; but that way of putting it lends itself to a confusion of the linguistic act of describing with the character that is attributed to something by that act and thus to a confusion of ontology with language use. It is, of course, true that we also stand in many other relations to things in our environment that do not involve their showing themselves to, or being identified by, us at all. This is certainly the case for all those relations that are ordinarily described as causal. It is also true that if we *think* about such relations, it will be in terms of the *ti esti*—the whatness or quiddity—of the entities involved. But the latter do not have to be objects of either thought or perception in order to act upon us or, for that matter, on anything else. In the sphere of perception and memory as well as of imagination and action which we think of as being most quintessentially human, things *are* identified and understood in terms of what they are. Indeed, if, in one sense of that much controverted word, the *being* of things consists in the characters they bear, in another it is just the fact of their presence in the sense of their having a place in an open domain—a world—that makes them available for those beings who may be said to *have* a world.

It does not follow from what has just been said that things present themselves to us in just one way which therefore constitutes what these things must, in some non-optional manner, be identified as being. Essentialism of that kind has no part to play in the position I am outlining here. It will also become evident that the way in which things do get identified has far more to do with pragmatic considerations than with eidetic intuitions of any kind. What is important here is the idea that, instead of having to deal with the mental proxies of things in the world, we are dealing, without intermediaries, with those things in the world themselves. We are doing so, however, in a way that involves selection and selective emphasis in the way we determine what they are. Those things are for us what they have been identified as being in some context in which we and others have encountered them. We deal with them, not by consuming them or incorporating them into our bodies or our minds, but at a perceptual and semantic distance and in terms of some character that they bear. This last is a point that Santayana often made; and it has been even better expressed by saying that "we touch these objects only with respect."[4]

Since the mode of presence of things to us is at issue here, it may be just as well to remind ourselves of a number of truisms that have a bearing on this relationship. The relation in which we stand to things, for example, is not reciprocal. They are there for us, but we are not, for the most part, there for them, the exception being, of course, other human beings and at least some animals. Nor are they present to themselves as we are to ourselves. There is no possibility, therefore, of there being either a difference or an identity between the way in which they are present to themselves and the way they are present to us. It follows that nothing of that kind can be in any way a standard for our ways of identifying things in the world. This does not mean that there are no constraints on the way we construe these identities, but they are those of practical utility and of empirical convenience as, for example, when we treat what moves as a unit as one thing.

All of this is so familiar as to seem almost too banal to warrant being repeated. What else do you expect, it might be asked, and what could things be if not things of some kind or other? This response is not surprising since once entities like us are in being and, as one might put it, in business, dealing in identities and meanings, it becomes absolutely natural for us to think of everything in the world in those terms. It is presumably this fact that accounts for the indelible strain of Platonism that marks any mode of thought that concerns itself, not so much with what makes things happen, as with what things are. Even when we think of a world in which there might be no one capable of this kind of identifying reference to things, we think of it in the same way and supply the standpoint from which these identifications can be effected. The point that needs to be made about this is not that if there were really "no one around in the Quad," things would no longer be what they otherwise are and thus, in effect, lose their identities. What would go missing is rather any form of life in which things are dealt with on the basis of what they are.

When there is such a form of life, we are dealing with states of affairs—something's being such-and-such—rather than with things as bare particulars.

We have, in short, entered a world of meaning and of relations of meaning in which the fact that something is a such-and-such can figure in our dealings with it in a variety of ways. It can, for example, be the reason why something can be assumed to be the case or why something should be done. Some philosophers have spoken of a contrast between a "space of causes" and a "space of reasons"; and that contrast is very relevant here.[5] What I am suggesting is that presence and the kind of world that presence makes possible is a necessary condition for a space of reasons. There really is no difference between such a space and "the world" if it is understood that "world" here has the sense of what we are in and not simply that of an aggregate of entities. In such a space, even the causal character of the processes that go on within the world has to be rendered in terms of the "reasons" why something happens even though reasons as such have no place in and no influence over the space of causes. To this, it must be added that this world as a space of reasons is also a "realm of truth"—a world in which questions can be asked about *what* something is and attempts can be made to answer this question.

The more general point I want to make here is that truth and all the matters just touched upon that in one way or another are bound up with it presuppose presence. This is to say that truth itself—things being one way rather than another—involves an entity or kind of entity *for* which things can be what they are. This claim goes against established conceptions of objectivity that separate truth very strictly from anything that smacks of the psychological or the subjective. The currently favored way of doing this is to treat truth as a property of statements and of statements conceived in purely logical terms and in the most complete isolation from anything having to do with the person who makes them and any motives or intentions that may have been associated with their utterance. In this way the truth-relation is supposed to be maximally objective since it holds between a state of affairs in the world and a statement conceived as a form of words—that is, in a way that treats it to the extent possible as though it, too, were a relation between objects. Even the match between the two—the statement and what it is about—is considered in abstraction from the involvement in it of any being for which the one and the other would be available for the purpose of making this comparison.

Plainly, though, these precautions against a contamination of truth by contact with psychological matters have no relevance when the whole idea of immanence in a mind has been avoided as rigorously as it has here. Presence is not a psychological concept and its function is simply to draw attention to what has been implicit in even the most austerely anti-psychological conceptions of the word–world relation. The idea that words can be *about* something in the world even though that something never shows itself or appears is the product of a form of abstraction from the real context of inquiry that cannot be sustained. If taken at face value, it would entail the thesis that the match (or mis-match) of a proposition with a state of affairs is fully independent of anything that could be called a comparison and that, as such, would require the presence of the two terms to someone. It may well be that this two-term conception of truth is itself open to objections on the grounds of its hypostatizing the propositional term of comparison. What is clear is that truth in its most fundamental sense is the self-manifesting of the thing it concerns to someone—someone who either has asked or can ask a question about it. Anything we inquire into is what it shows itself as being whether or not we are able to pose the right questions to it or to set what we see in the right kind of context. Its being, in other words, is not dependent on our questions or our identifications and its self-manifestation is simply another name for the presence that is the constitutive feature of perception.

These very abstruse ontological matters can be shown to have bearing on logical issues that may make them more concrete and manageable. It is sometimes thought that the truth-value of what we say about things in our world should ideally be independent of the character in terms of which these things have been identified in such statements. As long as what is described in one way is the same thing as what is described in another, the truth or falsity of the statement that uses the former description should carry over to any other that says the same thing about the object in question. This would mean that if a sentence like "John knows that Socrates died by drinking hemlock" is true, then another sentence like "John knows that the husband of Xantippe died by drinking hemlock" must also be true even if John does not know that Socrates was the husband of Xantippe. In this way, language and the truth conditions for sentences are effectively separated from the human beings who frame such sentences.

This view may well satisfy our sense that if on these occasions we are saying the same thing about the same thing, it would be perverse to treat one such statement as being true and the other not. Nevertheless, the consequences of achieving "objectivity" in this way must not be taken lightly. The terms of reference we use invariably identify the things we talk about in ways that express the character that has been manifest in someone's commerce with those things. Doubtless, the latter also have other characters in terms of which they could be envisaged in statements about them, including statements like the one in my example. The fact that Socrates was the same person as the husband of Xantippe may be uncontroversially true for those who formulate this example; but it simply does not follow that this truth makes sentences about what other people know about the fate of the husband of Xantippe true as well. The demand that they should be true seems to presuppose that all truth conditions are set by a "view from nowhere" and in a way that simply sets aside the fact of the plurality of beings who can think in terms of truth and falsity. Unfortunately, such an assumption has the effect of requiring every "view from somewhere" to deny its own indefeasible reality. It is, after all, one thing to accept the truth of "John knows that the husband of Xantippe died by drinking hemlock" once John knows that Socrates and the husband of Xantippe are one and the same person. To insist that it must be true independently of any such discovery is to make truth independent of the plural character of the beings who are in a position to recognize it. It is also, finally, to separate truth completely from the presence in which something can *be* something, either for itself or for someone else.

The real point here is that if sentences that are not truth-functional are effectively eliminated from the language of science, it will be impossible to express in words the fact that there are "subject-entities" in the world with other entities. It does not seem too much to say that the purpose of any such program in the philosophy of science is to sever the puzzling linkage that belief is supposed to effect between two entities like the "John" and "Socrates" of my example. In the kinds of revisionary locutions that have been proposed, anything we want to say about the one and/or the other will have to be cast in the form of a paratactic construction in which two quite distinct sentences, like the corresponding facts, stand on their own and quite separately from one another. With this modification of the original sentence, there will not even be a hint to suggest that there is anything or anyone in the world that stands to either Socrates or John in any relation other than those that are external enough to be expressed in this format and thus to have the official approval of the scientific world-view.

It may be appropriate to take note here of a tendency of contemporary thought that may seem to go in a different direction from the one just discussed. I have in mind the strong emphasis that has recently been placed on the fluidity and impermanence of the lines that separate one entity from another and, in so doing, divide the world up into different entities and different kinds of entity. Instead of claiming that these demarcations are laid down *in rerum natura* we now prefer to view them as creatures of language and culture and as being validated, finally, by considerations of pragmatic utility. All this is true, but the fact remains that once we have committed ourselves to a particular way of classifying and individuating what we find around us in the world, there appears to be very little willingness to conceive the relations among the entities so constituted otherwise than in terms of spatio-temporal position and causality. Bishop Butler's maxim that "each thing is what it is and not another thing" has sunk in very deeply, it seems—so deeply that it blocks any way of conceiving the relations among entities that would tend to call into question their separateness from one another. As applied to us, this would mean that what we are must be specifiable in a way that does not uneliminably involve any relation to other entities. And yet, in this version of ourselves, we are portrayed as giving names to things in a manner that expresses our own central capabilities—in psychological parlance, those of thought and perception and speech. But if these modes of functioning and all the others that depend on them cannot be described without stepping outside the logical boundaries that are supposed to make us just the one thing that each of us is, then the attribution of this kind of separateness to human beings must be a fundamental error. It is an error because human beings are, in a special way, together with other entities before they are put apart by the application to them of Butler's maxim.

4
THE SUBSTITUTION OF LANGUAGE FOR PRESENCE
Or words as things

I

Although language occupies a position of special importance in contemporary philosophy, this "linguistic turn" need not by itself predispose those who make it in favor of naturalism. Indeed, a primary interest in the kinds of logical issues that occupy linguistic philosophers tends to have just the opposite effect. There is, however, an area in which the concerns of linguistic philosophy and naturalism intersect. Traditionally, philosophers interested in the mind and mental functioning have tended to think that it should be possible largely to bypass the utterances in which mental functions are expressed and to concentrate attention directly on what takes place in the mind when we perceive or remember or whatever. Over time, however, dissatisfaction with the fruits of such inquiries became widespread among both philosophers and psychologists. Questions were raised and doubts expressed about the reliability of the kind of introspection such inquiries appeared to rely on. In contrast to the private character of most mental functions, the great advantage of language and language use was thought to be their accessibility. What someone says, unlike what he thinks, is not hidden from public view; it is expressed in words that anyone can hear or, if they are written, see. As such a public fact, moreover, language lends itself to joint, cooperative inquiry as introspection hardly does. On the strength of considerations of this kind, language has come to occupy a privileged place among the topics that come under the general rubric of the mental. At times, it has almost seemed as though the language we use for talking about mental functions were altogether displacing those functions themselves. It was, therefore, quite in the spirit of these views about language for a philosopher to propose, as one did not so long ago, that dreams be simply dismissed as mental episodes and that the stories people tell when they wake up be substituted for them.[1] As speech or writing, these stories apparently were not thought to be problematic in a philosophical way, as dreams are.

The especially attractive feature of language use is the fact that, in addition to being publicly accessible, it has a semantic character that other forms of human behavior do not have. This makes it possible to apply the concepts of truth and falsity to what is said as can hardly be done in the case of bodily processes like the digestion of food. Speech as a form of behavior is thus not only a respectable object of scientific inquiry, it is also a vehicle of truth. As such, it performs an essential function of "mind" and it does so without generating any of the puzzles usually associated with mental functions. This makes it hardly surprising that an approach to the philosophy of mind through language and speech has come to be regarded as a way of having the best of both worlds and the disadvantages of neither.

Plainly, though, a great deal depends on the way language itself is conceived; and this is where the concerns of naturalism and linguistic philosophy intersect. Linguistic philosophy wanted something to work with that was less elusive and impalpable than mental states were reputed to be. Even so, the difficulty about this was not conceived in ontological terms and did not presuppose any special scientific or philosophical view of what words are. By contrast, the adherents of hard naturalism appear simply to have assumed that language can only be verbal behavior and that for this reason it can be smoothly integrated into the same kind of physical process as any other natural phenomenon. Now when inquiry into language remains largely unencumbered by any special scientific baggage or the jargon associated with it, it may seem as though there were an affinity between this kind of linguistic philosophy and the natural attitude. No interest in such an affinity has been shown by linguistic philosophers, however, and most especially not in the idea that language itself raises ontological issues. As a result, the relation in which this kind of philosophy stands to scientific theory has never been clearly defined and the broader relation of the world of the natural attitude to the scientific world-view has been left largely unexplored. Perhaps this is why this kind of ordinary language approach did not have much in the way of instruments of self-defense when the sharp turn toward a science-based philosophy that was described earlier took place, and tended to give ground rather quickly in the face of its science-based opposition.

In spite of these important differences, a version of the pre-eminence of language as the primary datum for philosophical work has maintained itself through the movement of philosophy toward a closer relation to the natural sciences. Although the status of language is no longer defined by the prescientific assumptions of common sense, there still appears to be a widely shared conviction that language, perhaps alone, has survived the wreck of all the old mentalistic apparatus of philosophy. What is thus presupposed is that language presents no serious obstacle to its integration, at least in principle, into a naturalistic ontology where it would be on the same footing, epistemically and ontologically, with the other (physical) facts of the situations under study. More concretely, that means that the status of the statements that ostensibly report various mental goings-on have to be interpreted in accordance with the naturalistic assumptions of the inquiry in which they figure. It is assumed that the public character of language makes it a physical phenomenon that satisfies the conditions set by hard naturalism for the objects of scientific study. Language as speech is thus effectively substituted for mental fact.

It is the conception of speech and of language that results from this drastic *Gleichschaltung* that I want to examine critically in this chapter. What appeals most strongly to the naturalist is the fact that, by virtue of its consisting of sounds and marks that all can observe, speech is reassuringly concrete as are the movements of the vocal chords and the hand that produce these sounds and marks. Against this line of thought, I will argue that even if it prevails against traditional dualism, the de-mentalization of language that naturalism effects has grave difficulties of its own. These result from the assumption that a satisfactory account of language and language use can be given in the terms made available by a physicalistic ontology. Within such an ontology, the relations among speakers, words, and the things that these words are usually said to stand for would all be assigned to the same level of observable physical (and mostly behavioral) events. I will try to show that although these relations are not relations of presence as in the case of perception, they do nevertheless involve a variant of presence that I will call "presence in absence." In a sense, therefore, the anti-dualistic argument that naturalism mounts is irrelevant and what counts against naturalism is the fact that it attempts to transplant language into an ontological milieu in which it cannot survive.

II

In Chapter 2 it was pointed out that our understanding of perception tends to center on our sense organs and that this has the effect of making it appear that perception, generally, is something that takes place in our bodies. The real achievement of perception, I argued, would have to be conceived quite differently in terms of a distinctive way of being with things that are (mostly) not parts of our bodies. Something similar can be observed in the case of language. Here the "misplaced concreteness" I spoke of in connection with perception accrues to utterance as such—the production of words and sentences by the organs of speech. Here, too, the consequence can be a distorted understanding of language as a human function and even of speech itself. As before, the distortion has to do with the role of things other than speech itself. More concretely, it makes it appear that speech is rather one-sidedly the activity of producing utterances and it thereby obscures the way language is tied to the world which it articulates by dividing it up into distinct entities and kinds of entities.

Linked to this tendency is a way of treating words as though they were things—"word-things," in Heidegger's phrase—and, as such, on a dead level with all the other things that make up the world. Paradoxically, this way of conceiving words stems from a failure to give full value to the concept of the world for which the concept of nature as understood in the natural sciences is substituted. Oddly enough, a phrase J.L. Austin—hardly a physicalist!—once used expresses very well, though no doubt unintentionally, what is involved here. He said that, in order to explicate the logical functions of the words it studies, the philosophy of language has to "prise words off the world." What he meant by this phrase was probably akin to what Husserl had in mind in speaking of "bracketing" the existential assertions that express the natural attitude. By analogy, we have to stand back from our own use of words to designate the actual situations we deal with so that we can examine how these words and expressions function. The question about this is whether in prising these words off the world we also try to abstract from the fact that, although words are not things, they are, as one might put it, inherently "worldly." This means that, taken in isolation from the world, they are just sounds and marks, not words. Paradoxically, it is precisely this worldly character of words that is missed when they are treated in this way. This is what I want to explain in this section.

The thesis I will be defending is that the nature of speech (and thus of language) is radically misconceived when it is treated as somehow compatible with the underlying assumptions of naturalism as other mental functions are thought not to be, at least in our customary understanding of them. An initial observation one might make about this claim is that it is guilty of an *ignoratio elenchi*. Quite simply, speech in any real-life situation involves—indeed, it requires—perception. When language is spoken of as "public" and in an implied contrast with what is sequestered from view or from the view of everyone except one person, it is being treated as something we can all perceive and consequently as something that has its place in a common world. The person to whom we speak has to hear the words we speak and we have to hear what he says in reply. But if speech is tied to perception in this way, it would not be difficult to make a similar case for the essential role of memory in language use and probably for some other—traditionally "mental"—functions as well. In that case, however, it makes sense to assume that speech belongs to the same ontological context as perception and that it resists reduction to behavior just as perception has been shown to do. This context that speech and language use presuppose is not that of the physical world as this concept is construed by the sciences of nature. This is not because speech belongs somewhere else rather than to the familiar world of everyday experience, but because the character of that world that makes language possible is erased when the nature of physics replaces it.

Speech is, in fact, the most obvious counterexample to the claim that what is public must be physical and that the world of our experience is pervasively physical in some sense that brings it under the concept of matter deployed by the natural sciences. This should not be misunderstood as implying that speech has no physical character at all. As will become increasingly apparent, what is non-physical about speech as well as the other functions under discussion here is the fact that it embeds its physical elements in a relational context of a kind for which the natural world itself offers no plausible analogies. Sometimes this point or something like it appears to be half-way acknowledged by those who argue that a philosophy of language presupposes a philosophy of mind. It is doubtful, however, whether such acknowledgments have led to any deeper consideration of the ontological status of language; and the only "philosophy of mind" that has come over the horizon lately has been hard naturalism. Certainly there has been no willingness to face up to the detailed implications of what naturalism means for our whole understanding of speech and language.

One such implication is that reference as the relation between a word and the thing it stands for would have to be understood simply as the relation between two physical objects or events. In fact, however, none of the physical relations in which these "objects"—the sound and the thing—stand to one another could be used to pick out the ones that actually have a referential character from those that do not. In our usual way of thinking about such matters, words are supposed to have a meaning that ties them to the objects they denote. That sort of nexus is ruled out by hard naturalism so the only way it can try to provide some linkage between these two terms is by showing that the behavior of the user of words is the key to what we think of as their meaning. It is true that in many situations we name something that we are also handling or using in some way; and in such cases our behavior might well serve as a clue to what our words mean if such a clue were needed. It does not seem, however, that this explanation can cast any light on the way the person who speaks himself understands the words he uses. He does not wait around, so to speak, and observe his own behavior and extract the meaning of his own words from those observations.

Once again, what such an approach comes to is the idea that the meaning of what someone says must finally be in the hands of someone else—an observer who, one suspects, may turn out to be a behavioral scientist. He is supposed to be able to set the sounds and marks we laymen produce in the context of a pattern of conduct that assigns a meaning and a reference to them. But just as happened in connection with a similar treatment of perception, there is a difficulty here and it is the same one that arose there. The scientific inquirer who objectifies other people in this fashion does not and cannot do the same thing in his own case. He does not and cannot treat what he says about, for example, the language use of someone else as being interpretable only in the wider context of his own behavior. If what he says in this capacity were interpretable only in that way, he would not know what he had said until he had watched himself react to a statement he himself had made but did not yet understand. But why would he (and how could he) react in any way at all when he does not yet understand its import? In actual fact, there is no need for such self-observation since the investigator already knows what he has said because he already knows the meanings of the words he

THE SUBSTITUTION OF LANGUAGE FOR PRESENCE

uses. What is more, this is presupposed by the very explanatory "theory" on which he is relying so confidently.

Nevertheless, there are undoubtedly those among the sponsors of such theories who are prepared to take the heroic course of immolating themselves on the altar of their theory. This means rejecting any special or privileged position for themselves and insisting that the theory applies to them just as it does to the people they observe. Unfortunately, it can be shown that, if they take this line, that theory will no longer be, in any meaningful sense, *their* theory. This is because, by so doing, they will have placed themselves in the object domain of that theory. That can be unobjectionable as it was in the case of Newton who was subject to the law of gravity that he himself discovered. There is an important difference, however, between the law of gravity (and other physical laws) and this conception of language. The difference is that the latter is a special application of a philosophical principle—that of physicalism—which denies that anything is ever present in the sense of showing itself to someone. The theorist of language who accepts that his physicalistic theory of language applies to him will stand in a relation to his own theory that is the same as the one in which he stands to the functioning of his liver or his kidneys. He will, in other words, be quite without access to what is asserted in that theory as well as to the observations on which it is founded and by which it is supposed to be confirmed. He may utter words and sentences that would ordinarily be taken to express what the theory asserts; but, by his own account, these utterances will be purely physical events and, as such, parts of or events in his neural system and accessible to him only via their behavioral sequels. Not only is this last idea deeply incongruous, but there is every reason to doubt whether the behavior so observed would be interpretable with any accuracy.

III

In speech we do not just give names to things. We also say things about them. We say, for example, "The snow is white"; and it has been famously pointed out that this statement is true only if the snow is white. What is said in this statement—that the snow is white—is what we call a fact or a state of affairs. It is also this state of affairs, we would ordinarily say, that makes the statement, "The snow is white," true. But if facts or states of affairs play such an important role in connection with truth and cognition, they also present a grave difficulty for the naturalist. Where, after all, are such facts to be accommodated in a world that is conceived as an aggregate of things? A thing by itself is certainly not a state of affairs; and, unlike a state of affairs, it cannot make anything true or false. It can do that only if it is itself embedded in a state of affairs. If, for example, it is right there in front of me, we have a state of affairs about which we could make statements that might be true or false. But a state of affairs simply does not fit into the kind of slot in the world that

THE SUBSTITUTION OF LANGUAGE FOR PRESENCE

a thing occupies; and if the world is a totality of things as naturalism insists it is, what is naturalism to make of states of affairs?

There has been a strong disposition on the part of many philosophers to resolve the difficulty by assigning states of affairs to language. It is hard to say whether this idea has any special naturalistic credentials; but it does seem to reflect the view of language as somehow concrete and publicly available that has just been described. However that may be, it is clear that if naturalists espouse this view, it will present them with a difficult choice that bears on their way of conceiving language itself. The difficulty consists in the fact that if states of affairs are in some sense the work of language and as such are not in the world in the way that things are, a naturalistic conception of language in terms of sounds and marks and behavior will be pretty well ruled out. The reason is that when language is understood in that way, it simply adds a new set of objects to the world—vocal sounds, ink marks, and the movements that produce these. Understood just as things, these do not bring any states of affairs into being and consequently do not need to be expelled from the world. These new objects would stand in the same relations to one another and to the objects already in the world as the latter do to one another. Those relations are spatio-temporal and sometimes causal; they are not syntactical relations of the kind that obtain between the elements of states of affairs as expressed in sentences nor are they the logical relations that hold among different states of affairs. The latter center on the notion of consistency: some states of affairs can "stand with" certain others and still others cannot. It cannot, for example, be the case that all men are mortal if John Jones is not mortal; but John Jones and Socrates can both be mortal without any difficulty. No provision is made for such relations among facts by hard naturalism; and language as a supplementary set of objects in the world certainly cannot express anything of this kind.

Quite apart from this difficulty, the claim that states of affairs are not in the world is hard to understand if it is supposed to mean more than that states of affairs are not things. The idea seems to be that because states of affairs are not things, they cannot be in the world with things. On the other hand, separating the two from one another by taking states of affairs out of the world does not sound at all promising, if it is even intelligible, since without things there could be no states of affairs. The picture here seems to be one in which language as well as the beings who use it are intruders in a world of things and proceed to invent something—states of affairs—that is really not a part of the world at all. These states must accordingly be viewed as pertaining exclusively to language and to the beings—ourselves—who produce and use language. But one has to ask: how does this assigning to us of what appears to belong to the world really differ from the distinction that dualism used to make between what is really in the world and what is in the mind? There are not too many places where one can put something that is taken out of the world and the mind has long been the favored venue for such transfers.

But if the conclusions reached in earlier chapters about assigning things to the mind are sound, then banishing states of affairs from the world has to look very much like a *pis-aller*.

There simply is no good reason for thinking that things and states of affairs cannot be in the world together. Indeed, there is every reason to think that they must be in the world together although they are in it in very different ways. A state of affairs is not just another item in the totality of things that make up the world. It is in the world as what constitutes the openness of the world that is presupposed in everything we say and claim to know about those things. What this means is that if I imagine the most rudimentary apprehension I can have—something's simply being there in front of me—the only way I have of expressing this is the kind of locution that says that something is the case and thus expresses a state of affairs. I am, in fact, in the world as that to which some things in the world are present; but *I* am not present to them unless they are entities like me. In this sense, both truth and falsity as well as the states of affairs that make something true or false are grounded in presence—the openness of the world—as a necessary condition. If it were not the case that things show themselves and are there for us and for anyone like us, there would be no states of affairs inside or outside the world and no way in which things could figure in true (or false) statements about them.

Although the idea of states of affairs being in language and not in the world has little to recommend it as it stands, it is possible that the intuition that underlies it is sound. It insists on a point that is both valid and important—the special status of states of affairs—and it goes wrong principally by virtue of trying to do justice to the difference between such states and things in a dualistic manner—that is, by taking the former out of the world altogether. This seems to reflect the conviction that something must either be in the world in the manner of things or not be in it at all. This not only confuses "the world" with "nature"; it also altogether misses the prior question about how we who are the bearers of language and thus, on this view, of states of affairs as well, are in the world ourselves. If we are in it not in the manner of things but as the entities for which other entities are there, then states of affairs become the indices of this mode of being and the language in which they are expressed is one principal modality in which things in the world are disclosed.

IV

This idea of disclosure is common to the discussion of perception in Chapter 2 and that of language here. It derives from the most fundamental understanding we have of ourselves as beings to whom entities other than ourselves as well as we ourselves are present in the various modalities of presence. These are also the entities that, in the case of language, we talk about ourselves and learn about from the talk of other people. In the account that was given here of perception, it was argued that views that assert a purely external and causal connection between perception and its objects cannot render this fact of disclosure. What has been said about language and speech up to this point clearly indicates that naturalism imposes a similar treatment on them. It makes the relation of a word to what it is supposed to designate an external one in just the way that of perception to its object is held to be. In order to counteract this line of thought, language itself has to be shown to have a disclosive character that stands in the way of any attempt to assimilate it to a model of thing-to-thing relations.

It is a good deal harder to make this claim plausible in the case of the relation of words to things than it was in that of perception. We have become accustomed to thinking of words as having a distinct reality of their own that makes it seem easy to contrast them with the things they stand for. In perception, by contrast, although "sensations" have been supposed to be the counterpart to words for the purpose of a similar contrast, the object that is perceived dominates the perceptual situation so completely that it leaves little room for a two-term contrast along these lines. But in spite of the opacity we attribute to words as unit-entities on their own, language can be shown to have a disclosive function that is different from but also akin to that of perception.

The question is how the appearance of an external relation between words and what they denote can be set aside without any appeals to a magical coalescence of word and thing. I think this question can be best approached via the fact that we are able to speak (and speak to one another) of things that are not perceptually present. In such cases the distinctness of the words being spoken—the materiality or sound-character of these words—from the things to which they refer is especially clear since the thing in question is not perceptually present at all. Even so, the sound of the word is absorbed into its signifying function and takes on the "physiognomy" of the thing it is used to name, to the point where we may hardly be aware of it as a sound. This fact can be acknowledged by naturalism only as some kind of psychological accompaniment of language use that has no bearing on the ontological status of words. And yet this approach to language would make the meaning of words just such an external accompaniment of something—a sound—that as such has no claim to be regarded as a word.

In the discussion of perception it was argued that what has been called "naive realism" has a much greater validity than it is usually thought to have. In the case of language, a similar argument can be made that would call into question the idea that the relation between words as sounds and the things that words refer to is one of mutual externality and that each term in the relation can accordingly be described in full independence from the other. This does not mean that speech summons up the things it signifies so that they will be present to us as they are in perception. What it does mean is that a word would not be a word but merely a sound if it did not *name* something other than itself. Being a name is not a property of a sound in any way that a scientific inquiry could reveal. Nor is it a feeling on the part of the person who uses a word or any other mental or behavioral accompaniment of the use of that word. What makes a word a word is the fact that the speaker is in the world in which the thing designated by the word is to be found. The word is, in fact, one form that the presence to the speaker of that world assumes. It is also the presence, putatively this time, of whatever it is in the world that the word denotes. If the presence of the world is undeniable, that of any particular thing in it cannot be guaranteed. What makes a word a word is not therefore some magical link to a particular thing—a link that would guarantee the existence of that thing. It is the fact that the word refers to the world in which that thing must be found if it exists at all.

We have all heard that there are or have been people—usually described as "primitive"—who were/are unable to make the distinction between words and things. The suggestion seems to be that such people believe that some material coalescence or fusion of word and thing occurs. In the light of what has been said, however, it seems possible and even likely that there has been some confusion in the way the beliefs of such persons have been interpreted. It may be that they were saying something quite correct in the only way that was available to them. Perhaps it is not just a primitive delusion to think that word and thing should not be distinguished from one another in the way two things are. It may even be possible that, when language was still a relative novelty on the human scene, there was a sense, however crudely expressed, of what is distinctive of the relation that does hold between a word and the thing it stands for. Doubtless, that understanding, if such it was, was distorted by the analogies to the physical processes of coalescence in terms of which it seems to have been described. Today, we are so put off by the idea of there being something magical about the words that make reference to things possible that we tend to take a needlessly hard line against anything that sounds to us like a departure from the approved doctrine. It may be the case, however, that we react in this way because we have lost any sense of the respects in which language really does resist assimilation to the categories of scientific thought. We remain tenaciously committed to the idea that language, even in its most problematic aspects like meaning, must eventually yield to an approach based on the categories of the scientific world-view.

If, against the background of such considerations as these, one asks what a word is, the only satisfactory answer will be that it is, not the thing itself that it stands for, but the presence to us of that thing which may or may not be perceptually present itself. This is typically a presence in absence of the kind we are all familiar with from the experience of, as it is usually expressed, "thinking" of something that is not perceptually present to us. We usually leave such references to something absent completely unexplained or try to spell out what they involve by invoking imagery or some other psychological process. The alternative is to recognize words themselves as entities of an altogether distinctive kind in which presence and absence are combined as they are quite generally in the lives of human beings. Words are, of course, sounds as well; and the distinction between the thing referred to itself and the materiality or sound-character of the word remains intact. But, in the semantic function that constitutes it as a word, a word is the putative presence (or presence in absence) of the thing it is said to stand for.

This is the kind of relation that philosophers call intentional; and as far as it goes, this usage is unobjectionable. In one respect, however, it can be quite misleading. The associations most people have with this notion are such as to make it something that goes on in the mind—our beliefs, for example, simply as beliefs—rather than a *bona fide* relation between two entities. The term I would propose in its place is "transcendence," understood as the way in which we are able to refer to an entity that may not be perceptually present to us but is emphatically that with which we are concerned in some way or other. Transcendence has this much in common with intentionality: in both cases the thing with which we are so concerned may not exist at all. In the case of intentionality, however, this possibility is built into the concept itself in such a way that its *only* use is to designate the mental side of this outreach. By contrast, the paradigm case for transcendence is the one in which both terms of the relation are real. This is also the presumption that guides our understanding of language although it has never been properly given its place in philosophical accounts of language.

It is also clear, for reasons that have already been explained, that to speak in these contexts of "access" (or the lack thereof) is really to invoke a notion that has no place in the naturalistic scheme of things because it names a relation that is not acknowledged there. It appears, however, that this very fact is the reason why any such theory issues in paradoxes of the kind just reviewed. It looks as though the source of the trouble here may be a failure to take seriously the distinction between a human being as a whole and its parts or organs that play a role in its various functions. One thing, at least, in all this should be beyond question; and that is that the views or beliefs someone holds are not physical parts of or processes in that person understood as an organism. They are assignable only to me as a whole human being and in any normal state of affairs they cannot be inaccessible to me in the way processes in my brain are. The idea of our having to scrutinize our own brains in order to learn, *per impossibile*, what we think, is utterly grotesque even if we imagine that the material difficulties in the way of our doing so could be removed. It is true that we sometimes don't know what we think; but in such cases a resort to the brain will be of no avail. Typically, this sort of thing occurs when we fail to recognize that a certain belief is implicit in what we say or do. In such cases, however, we can work out what we have been assuming and thereby come face to face with what we actually believe and acknowledge it as such. There is no possibility, however, of anything comparable when our vis-à-vis is our own brain.

To return to the idea of "prising words off the world," there can be no doubt that we have learned to do this and to recognize them as words, not things. And yet the question remains: what is it that we recognize them as being when we recognize them as words? It does not appear that naturalism has any half-way satisfactory answer to this question. It can only insist that they are just sounds or marks and that any other fact about them—any mode in which they may function or be used—is just as distinct from them as an arrow-like pattern of lines on a road-sign is from the direction in which we say it "points." But if that is really so, why is it that, in our dealings with them, words never do reduce to sounds and marks? As Heidegger pointed out, even when we hear people speak in a foreign language, what we hear is still not sounds but words—words we cannot understand. And what about what has been called the physiognomic character of words—the way in which they take on the atmosphere or coloring of the thing they name? In a general way, it seems much more plausible to say that although we can isolate the mark or sound that functions as a word, this remains something we have to *learn* to do. Not only that, but, as Gadamer has pointed out, the language we use is still for the most part curiously invisible to us as language because words are so saturated by the things they name.[2]

V

There is an aspect of language about which almost nothing has been said up to this point. This is its communicative function. Through language, we can in principle share the world as it is disclosed to/by each of us with one another. Thus, although I have never been in Africa, it is possible for me to learn a great deal about it through communication in a variety of modes with others who have been there. It has already been pointed out that the distinctively human way of being in the world is that of encountering other entities in their being—that is, as what they are. But the same entities that are disclosed to me are also disclosed to other human beings and we usually live with one another in a large measure of agreement about what these entities are. We are also reciprocally dependent on one another for much of the knowledge we have of the world. In all these respects, we show ourselves to be familiar with *being*, both as the specific identity that is attributable to a given entity and as the general fact of the presence to us and to anyone of entities as the kinds of entities they are.

By virtue of the power of language to convey our thoughts and perceptions—our knowledge—to others, we stand in a relation to one another for which there are, once again, no real analogues in the natural world. What is even more interesting is the light this kind of reciprocity throws on the nature of presence itself. As it happens, a good deal of attention has been given by philosophers of a generally naturalistic persuasion—like G.H. Mead—to certain aspects of communicative reciprocity in the relations of human beings with one another. Their principal emphasis has fallen on the role played in human development by learning to reverse one's own perspective—that is, to imagine being in someone else's situation and seeing oneself from that point of view. In these accounts, however, there has been no suggestion that there was anything in these relations that was beyond the range of the categories of naturalism. But if presence *simpliciter* is beyond that range, as I have tried to show it is, then reciprocal presence surely must be so as well. In any case, the reciprocal character of presence is fundamental to many relations in which we stand to one another and of which language is often the principal bearer.[3]

In this matter of how presence is to be conceived, there has always been a temptation to treat it as though it were a form of illumination. Sometimes we take the further step of thinking of our own vision as though it were like the beam of a flashlight that plays over the objects around us and makes them visible. Like most metaphors, this one expresses something that does genuinely characterize a certain aspect of visual perception, namely, the control we are able to exercise over the direction of our gaze. It would seem that in tactile perception the hand that explores the contours of an object we cannot see does something comparable. (The way in which our eyes can trace the contours of some object is the visual equivalent of this kind of exploration.) The tendency of such metaphors is to make presence seem to be not only transitive in character, but also something that is under our control across the board—almost something we can turn off and on at will. But when I am the one who is present or may be present to someone else, this volitional aspect of presence is absent. This is not, I think, just a matter of its being transferred to the person who is or may be perceiving me. Quite apart from whether or not I am present to someone else at this moment, there is the fact that I am visible, that I am there in the same world with others and not just as a viewer but as someone who is or may be present to these others.

It is this fact—that perceiving is bound up with being perceived—that Merleau-Ponty expressed by calling it a chiasmus.[4] This is a literary trope designating the way the same words recur, with their order reversed, in a line of poetry like "Love's fire heats water, water cools not love." In just such a way, self and other are intimately involved with one another in any relation of a human being to another human being; both are at once present themselves and the beings to whom someone else is present. The fact that presence is reciprocal in this way places all of us in fundamentally the same relation to one another. It also sets up a dialectic among those who are present to one another as active beings who can make a difference, one way or another, in one another's lives.

What I am trying to show is that this is the kind of context in which language and speech are embedded and by which they are made possible. They are concrete realizations of reciprocal presence and the parties to linguistic exchanges are dwellers within a common milieu that no one owns and in which they are there for one another. For this purpose I would like to

THE SUBSTITUTION OF LANGUAGE FOR PRESENCE

introduce the expression, "a milieu of presence." This is the world as a space in which things show themselves as what they are (and sometimes as what they are not) and those beings—notably, we ourselves—to whom they are thus present also encounter one another and live in this reciprocal presence as their permanent mode of being. It is also the kind of space in which we can address one another and do so with the understanding that what we say will be heard by someone and a reply made that is, like our own utterance, addressed to someone who can hear it and respond to it. This is very different from anything that could take place between two machines that are connected to one another by electronic means so that they can "talk" to one another—that is, can simulate communicative exchanges like the ones that occur between human beings. The difference is that in the case of such machines everything that occurs on such occasions is anticipated in the "programs" that have been set up for them and so the "conversation" that takes place is really between their human designers or users. Taken in isolation from those beings, these machines are simply elements in one continuous physical process that is not mediated by the kind of reciprocity that involves the presence to one another of the parties to this communication.

If these claims seem outsized, we should bear in mind that the idea of such a milieu corresponds rather exactly to the way we in fact think about the world we are in. We do not conceive it as a world in which every entity is sealed up in itself so that it cannot be said to be there for or to *face* any other. Instead, the world is implicitly understood as a zone of openness in which things show themselves in a certain orientation to those entities that are not only susceptible of being present themselves, but can have other entities present to them. It is easy to confuse this generalized presence with various things it is not—the transparency of the atmosphere, for example. When these confusions have been cleared away, it becomes evident that there is nothing in the corpus of scientific or common-sense knowledge that can explain presence causally or otherwise. It is always, instead, among the most fundamental presuppositions of such knowledge although it is also implicitly denied by the ontologies espoused by the sciences of nature.

Even so, there is, in these circumstances, a great temptation to postulate some special power with which we are endowed and to suppose that it is this power that makes other entities present to its possessor even though we are quite unable to explain how this comes about. There is, moreover, no way in which this inference can be given any real empirical content and certainly no basis at all for any attribution of a transitive character (in any direction) to presence or for any idea that something emanating from us could make things visible or otherwise present to us. All we can really say is that, in a milieu of presence, entities, generally, show themselves to certain ones among them and that this fact has an impersonal as well as a personal character. In other words, it is something to which we are subject quite as much as it is something that can be said to emanate from us if indeed it can be said to do so at all.

Metaphors are of questionable value when we are so completely unable to supply non-metaphorical accounts of what they are supposed to express. Nevertheless, if they are permissible at all in these contexts, it might be best to say that it is as though human beings (and, as we will see, other living things as well although in very different modes) were the outcome of a transition from simply being entities among other entities within the world to being in the world as a milieu of presence. Once again, such a claim is likely to be dismissed as fanciful and the counterclaim made that we are and always have been living within the same objective world—the only real one—that is erroneously supposed to have *become* my/our world in such a transition. But, as before, the answer to this charge must be that the world we are really familiar with is the world we are *in* in the mode of disclosing other entities that are in it with us. It is the "objective" world understood simply as the aggregate of entities existing side-by-side with one another that is in fact the product of reflective thought.

VI

The concept of the world as a milieu of presence has been introduced and proposed as the matrix within which language becomes possible. Language itself has been presented as a mapping of the relations in terms of which human beings order this milieu as well as their relations with one another within it. In one way or another, this whole ordering presupposes presence; and the crucially important fact about presence for these purposes is that it is plural and fragmented. Every human life as a locus of presence is thereby set the problem of finding a way of living with other human beings on some basis that has a chance of lasting more than five minutes. That typically requires some form of cooperation and cooperation requires shared understandings of the situations in which something needs to be done and some measure of agreement about what needs doing. Since I am arguing that language takes its fundamental character from this ontological matrix in which it is embedded, it may be appropriate to give at least a brief account of the patterns of relatedness that emerge in such a setting. These are, by the way, the very patterns of normativity that find no place within the objectivistic theory of the world that naturalism represents.

The thesis I want to defend is that the idea of truth as the way things in the world really are founds a relationship among human beings on a set of priorities for their actions that would be jointly acceptable to all as being in their interest. Truth here has a twofold meaning: it is both truth as the way things are and what we try to get right, and it is truth as the openness or accessibility that has to be presupposed if the effort to grasp things as they really are is to make sense. The vital point here is that this truth that we seek has to be the same for everyone even though we may not all be interested in the same parts of it. This is what makes truth central to the dialectic of persons; and it would not be inappropriate to claim that what I have called a milieu of presence is *ipso facto* a milieu of truth. It follows that it must be a common presupposition of those who inhabit such a milieu that things in the world are, quite generally, open to the inquiries that seek to determine what their character is. A corollary to this would surely be that this character is not under anyone's proprietary control and that in fact each of us needs others in the determination of what is true. The openness to anyone of the entities that make up the world—their openness in the character they bear—is not therefore individuated in the sense of being tied to the "illumination" that particular human beings might be thought to bring to their commerce with things within the world. It is rather something that can be determined only with the cooperation of other human beings to whom, for these purposes, we stand in an implicit relation of partnership.

To put the matter as I have just done greatly simplifies a very complex set of relationships. As we all know, we do have a proprietary interest in what we take to be the truth of this or that matter; and we are very reluctant to submit our version of it to anyone else's judgment or to any independent authority. This attitude doubtless has its roots in the practical and appropriative character of our ways of understanding the environing world and the stake each of us has in his version of that world. Challenges to the implicit assumptions we have made about the way things are threaten to dismantle or at least to dislocate the routines by which we live our lives and can even impinge on our sense of our own identities. Sometimes these routines are predicated on priorities we attribute to certain ends over others; and a challenge to these can also be cast in the form of a question about whether it is really true that one interest is more important than or somehow superior to another. Controversies of this kind occur not just between individuals, but between competing groups and the intensity of the feelings engaged on both sides rises as the interests involved widen out to include whole communities.

At the same time as we try to hang on to our construals of the world and to make them prevail, we also want to claim that they are not true simply because they are ours. If their being ours made them true, we would have to concede that the fact that the views defended by our rivals are theirs makes them true as well. Sometimes, people seem to be leaning towards some such concession as, for example, when they say that something is true *for* them. Although it may not be spelled out, this presumably means that something quite different may hold true for other persons or groups than does for us. But no one can really be satisfied with this kind of crippling amendment to the understandings we have about what being true entails. To have any authority—any real "clout"—the application of "true" to something must hold good for everyone, however disputed it may be.[5] It is, in other words, how things really are that is in dispute, not our differing opinions about them. Sometimes we can find a referee who is sufficiently dispassionate and intelligent to be trusted to speak for this common truth and determine how things really are. Sometimes this is not possible. Even so, we are committed to the claim that there is a truth of the matter and that it is patent in the sense of being accessible in principle to anyone who is willing to look for it in the right place.

What has been said should make it possible to understand in what sense truth is a normative concept. When it is envisaged simply as the set of true statements or as the way things comprehensively are, this normative character may be obscured by the purely "factual" aspect of truth. But if the plural character—the sheer manyness—of the beings who are in a position to determine what is really the case is borne in mind, it becomes much easier to think of truth in terms of a kind of "ought." It is what we *ought* to accept and affirm because it accords best with the experience of large numbers of people who have had the opportunity to inquire into whatever is at issue. As the way things really are in the world, it supplies the only basis on which people can cooperate with one another and successfully accomplish what serves their needs.

These may well sound like empty pronouncements and yet they have quite definite implications for the way we conduct ourselves in our relations with our interlocutors. For one thing, the beliefs people hold are often skewed in a way that reflects their particular situations and points of view; and the bearing of such beliefs on those of others which may be similarly keyed to a particular point of view may be hard to determine. But if they are to have any authority for anyone else, these beliefs have to be sorted out in such a way that it becomes clear what all the parties are and are not saying. An ability and a willingness to make that effort is part of what the normative position of the truth means within the life of a community.

It has been suggested that in a human community truth has the status of a norm. Although objectivity is generally understood to be an emphatic and refined conception of truth, one of the most striking features of current conceptions of an objective world is the fact that values and norms have no place in it. The only reality that is assigned to them is that of attitudes held by individual human beings. These differ very widely; and when they do, nothing in the competence of natural science can prove that some attitude that does not prevail in a given place at a given time ought to be preferred to those that do. In this sense, relativism is the price that has to be paid for conceiving the world in accordance with the requirements of scientific objectivity. The world has to be understood entirely in terms of what actually is or has been or predictably will be the case; and neither what is merely possible nor what notionally ought to be the case has any place in it.

Efforts have been made by those who feel unable to accept these conclusions to find "properties" of one kind or another that can do duty as the bases for a rational justification for non-relativistic ethical attitudes. This is the idea that drives much of the talk about "values" that is so widespread at present.

Unfortunately, the concept of values came into being as a term of reference for what is found to be good or enjoyable or desirable by human beings. As such, what counts as a value becomes the creature of individual preferences and this makes the world of values as inconsistent with itself as those preferences themselves are. What is more serious is the fact that values have no power to generate an obligation or to bind us to any course of action that we are not drawn to by a sense of its being attractive to or enjoyable for us. Altogether, "the place of value in a world of fact" appears to be a designation for a conundrum created by the way the world is conceived; it is not something that it makes sense to search for in a world so conceived.

If, instead of looking for values, we examine more closely the relation in which human beings stand to one another, the prospect of finding something that has broader normative implications is likely to improve. It has already been suggested that that relation is mediated by the truth as the way things really are and that we are all committed, willy-nilly, to a common truth as something we ought to conform our beliefs to. What needs to be emphasized now is that in this way a relation among human beings is constituted that becomes the basis for a wider kind of normativity that radiates out into every sphere of human life. This is not to say that our conduct necessarily conforms to the requirements of that relation; we know that it does not. What it does mean is that when violations occur, the norm that is violated is one that no one can really just repudiate.

It is not hard to see what the sources of these other norms might be. Every human being has interests and judges events generally and actions especially—his own and those of others—by the way they affect these interests. Other people are also able to judge whether what they are doing is advantageous or detrimental for those who are affected by their actions. All of these judgments are, of course, subject to error. It may turn out that what I think is in my interest is not and what I think is not may be so. The same applies to the judgments of others about the effect of what they do on still other people. The vital fact here is that in a great many situations, there can be communication between agents and patients about such matters and so there is the possibility of correcting any misinformation on which our assumptions about the effects of our actions may have been based. Out of such communication there emerge, in any community or organized society, shared understandings about what constitutes a harm and what a benefit. The core of morality is simply the requirement that we be able to justify what we do that affects other people to those who are so affected in the light of all the relevant facts about it. "Justify" here means to show persuasively that some action is either indifferent from the standpoint of the interests of others or that its effect on those interests is as good or better than that of any other action that could be performed in a given set of circumstances. Such a justification claims to be valid in the sense that the assertions of fact on which it rests are true and the inferences drawn from them are reasonable. They are, therefore, as sound for the person or persons to whom they are proposed as they are for the one who proposes them.

People can live together without accepting this kind of justificatory relationship to one another if there is a preponderance of power on the part of some of them that makes it impossible for the others to resist the will of the masters. The Spartans and their Messenian helots are one example of a society ordered in this way. There are also societies in which people are nominally free to express their opinions about policies that affect their interests, but suffer a variety of informal penalties if they do so. The idea of morality is that of an association in which everyone is recognized as having the right to make decisions about his own well-being and to contribute to a public determination of the good for its members. To claim that something is in the interest of someone else when it manifestly is not is to fail to respect that person as someone who can in principle understand the tendency of events and of policies as they affect him as well as one can oneself. He or she is after all a dweller in a milieu of truth as we all are and must recognize one another to be.

5

WHAT DOES THE BRAIN DO?

I

Up to this point, the line of argument I have been developing has addressed themes that are, in considerable measure, peculiar to contemporary naturalism. They are, as has already been noted, not the sort of thing that preoccupied the minds of classical materialists like Democritus or Hobbes. Now, however, it is time to turn to an aspect of the naturalistic approach to the philosophy of mind that would probably have struck those thinkers as being completely in line with their own way of conceiving materialism. This is the effort to show that the mind itself is physical in nature, a part of the body. It is, in short, the brain and the brain as neuroscience describes it.

There has recently been a strong uptake of interest in the brain on the part of philosophers, undoubtedly in good part as a result of the progress that is being made by neuroscientists in exploring the anatomy and functional organization of that organ. This has raised hopes on the part of a number of philosophers that this work may have reached a point at which it can contribute something of value to the analysis of issues in the philosophy of mind. Interest in the potential relevance of neuroscientific inquiries has become quite widespread and has reached a point where a degree of familiarity with the physiology of the brain and with the current state of research into its functions is commonly thought to be a pre-requisite for philosophers hoping to work in what is now called "cognitive science."

I do not myself have this kind of familiarity with contemporary neuroscience or cognitive science and so this chapter will offer a distinctly external perspective on those fields of inquiry. My justification for trespassing on this territory lies in the fact that there is an issue of paramount importance to this whole enterprise that can fairly be posed by those who stand outside or on the margins of current developments. Plainly, any judgment on the relevance of brain research to philosophy and to what it has to say about human beings must rest on assumptions about what it is that the brain is supposed to do. Although it might seem that this is the most elementary question that could possibly be posed to researchers in this field and that an answer to it must be ready and waiting for those who pose it, that turns out not to be the case. There is no generally accepted answer to this question that has a clear bearing on this matter of relevance; and the ones that are given are anything but transparently valid. This is what I want to show in this chapter. I will also argue that as long as this situation holds, the pretty well universal conviction that we could not think or feel or perceive if we had no brain cannot be cited, as it so often is, as a knock-down argument for hard naturalism.

This state of affairs—our not being able to give an account of why brain function is a necessary condition for perception and thought—is deeply frustrating to many. Such an account would have to go beyond such brute facts as that severing the optic nerve causes blindness and a blow to the head, unconsciousness. Admittedly, what "beyond" means here can hardly be made clear without proposing something like a substantive theory of one's own. Even so, it does seem clear that what we learn about the brain and what we know about ourselves without the benefit of neurophysiology belong to such *prima facie* different orders of fact that their simply being empirically associated with one another cannot give any real understanding of how these first-person and third-person events are connected.

Matters are not made any easier by the fact that on the scientific side there is hardly any acknowledgment of this situation or of the need for a satisfactory resolution of the conceptual difficulties in which we are caught. Instead, it seems to be simply assumed that all significant questions in this area of inquiry are scientific questions about the functioning of the brain and that answers to those questions can come only from the kind of research that is already in progress. At the same time, however, those who take this view borrow freely from the languages of common sense and sometimes even philosophy to describe the import of their work; and they do so without appearing to grasp the conceptual problems they are creating for themselves by stepping so nonchalantly outside the vocabularies of the sciences in which they work. It is as though it were simply being assumed that the rejection of dualism takes care of any philosophical complexities that might arise in connection with this work.

There is another fact that needs to be kept in mind when we deal with these issues. It has already been noted that nothing in perception or thought or feeling points to the brain as the site in which these functions occur. In fact, from an historical standpoint, the brain is only the most recent part of the body to be proposed as a locus for them. It is well known that other parts of the body like the heart and the *phrenes*—the lungs—have in the past been thought to be the seat of our intellectual and cognitive powers.[1] If such ideas now seem absurd to us, it is worth reflecting on the fact that they cannot be refuted by pointing to some internal feature of such functions by which the role in them of the brain is manifested. The point that needs to be made about this is that any satisfactory theory of the role of brain function in our lives should be able to account for the fact that if the brain makes it possible for us to think and perceive, it does so without ever putting in an appearance itself. Wittgenstein may have been pressing his point a little too hard when he said that, for all philosophical purposes, it would not make any difference if our heads were full of sawdust.[2] Even so, he did have a point; and although it can hardly be the final truth about this matter, we need to be able to understand how it could express even a part of that truth.

In what follows, then, I will be examining the answers that are given by naturalism to this question about what the brain does. It will turn out that these answers are at least partly borrowed from positions like dualism and even from the natural attitude and to that extent violate some of the central principles of hard naturalism. The result, as I will try to show, is a distinctly syncretistic account that seems to be unable to acknowledge its own patchwork character or its indebtedness to theories that naturalism otherwise repudiates. The thesis that brain function is a necessary condition for the functions we call "mental" is not in dispute. What I do challenge is the philosophical models of brain function in terms of which this thesis is currently interpreted.

II

The official position that naturalism takes on this matter of what the brain does is unabashedly physiological. It conceives the brain as a device that connects the neural input of the senses with behavioral output and does so without having to produce anything along the way other than the familiar electrical and chemical events that neuroscience regularly deals in. It may be doubted, however, whether even the most rigidly orthodox naturalist really hews to this line of thought with perfect consistency. There is, after all, the very widely shared idea that what goes on in the brain somehow produces effects that are usually described as being "in the mind" and, as such, are something we were familar with before we began thinking about these matters in scientific terms. As a claim about what is in the mind, this is quite obviously a dualistic conception and, as such, one that naturalism has pledged itself not to have any truck with. But when the brain is conceived simply as a device that connects the neural input of the senses with behavioral output, it is difficult even for many people who are well-disposed toward naturalism not to feel that something has been left out.

What has been left out can be variously described as our "mental state" or our "experience" or our "consciousness"; but it all comes down to pretty much the same thing. In the simplest terms, it is the fact that something *appears*, whether it be a sense-datum or a natural object or event. This is a fact we were already familiar with before we ever began to think about these matters in a scientific or philosophical way. If there is a sense of bafflement when the orthodox naturalistic doctrine about brain function is candidly stated, it may well be due to a disappointed expectation that this fact that we have learned to think about under various dualistic rubrics was to be shown to be causally dependent on the brain. Instead, it turns out to have been reduced to a *façon de parler*—the kind that is now often referred to condescendingly as a rather backward kind of "folk psychology." It expresses nothing that has any claim to reality, however dependent that reality might prove to be on the brain. It is an illusion that naturalism wants to sweep away and not something that needs explaining.

Although dualism and all its works have already been subjected to a severe critique and have been shown to be radically incompatible with the theses of naturalism, I want to examine these ideas about brain function that somehow fall flat when naturalism puts all its cards on the table. Briefly stated, what they come to can be described as a transmission theory. By "transmission" here I mean the conveying of information in either verbal or graphic form via some natural process like electric current. Telephones, television and radio are the most natural examples of transmission in this sense. In television, light that is bounced off some object reaches the camera and then, by electrical means, it reaches the eye where it sets in motion a neural process that reaches the brain. When we tell this story, there is almost a compulsion to go on to say that this process of transmission ends in something like a printout or a picture on a screen and that this is perceived and understood by some homunculus whom this message reaches. In this form, the message will convey to this being information about objects or events outside the brain. They are not perceived themselves, but their proxies on the inner screen are.

The appeal of this way of interpreting the function of the brain can hardly be overestimated. It expresses the sense we have of the way these electronic modalities of transmission work; and this understanding has been applied to our perceptual relation to the environing world as mediated by our sense organs and our nervous system. An interruption in the functioning of this system of transmission can effectively end our visual and auditory experience; and when anything of this kind occurs, it is as though a telephone line had gone down and some signal that would otherwise have reached us had been lost. It does not seem as though there could be any explanation of such facts as these otherwise than by reference to some physical change that has occurred in or around our own bodies and brains. It is very natural, therefore, to look for an explanation in some domain of scientific inquiry like neurophysiology.

There is another reason why the picture I have sketched of the mind in its relation to the brain seems so naturally right to us. Everything we know about the perceptual relation between ourselves and the things around us appears to indicate that there is a flow of stimuli from those objects to the sense organs of the perceiver but that nothing comparable goes in the opposite direction. It is true that in early speculation about what goes on in perception it was sometimes thought that there were two movements or "emanations" that made perception occur—one inbound from the perceptual object and the other outbound from the perceiver—and that perception occurred when these met. That idea evidently survives to this day, at least in the way children conceive vision as something that issues from the eyes. Needless to say, nothing of that kind has been discovered. As a result, the perceptual relation has come to be conceived as a one-way street and perception as something that is produced at one end of it. Such a picture as this can hardly be reconciled with any conception of the presence of the perceptual object itself to the perceiver. And if, as is commonly believed, the functioning of the brain and the central nervous system turns perceptual consciousness on and off, then the relation implicit in what I have been calling "having a world" cannot be a primitive one.

To this it must be added that all these considerations that speak in favor of dualism have been greatly reinforced by a development that especially affects those whose thinking has been shaped by the scientific world-view. For such persons, the natural attitude has quite simply become incomprehensible, at least at the level of theory, although they rely on it unquestioningly in other life contexts. The stone of offense here is the idea that something that is located at a distance from our own bodies should nevertheless be present to us as it appears to be in naive perception and not simply be represented by some reasonable facsimile. The conclusion to which this points is that, if our familiarity with things around us in the world is to be intelligible, they must quite literally be brought to us. This means that there must be some physical process that spans the gap between them and us and that process must, when it reaches our brains, bring into being some sort of replica of the thing we claim to see or otherwise know.

The argument I will be making is that the transmission theory works only insofar as it postulates a hookup between sensory stimulation and behavior or between incoming and outgoing neural activity in the brain. There simply is no evidence that this hookup includes anything like what we more or less automatically introduce into this picture—an "inner" screen with representations of external objects. The whole idea that any such "representation" takes place seems to be motivated by an attempt to reconcile the fact that something does appear with the considerations that militate against accepting the thesis of the natural attitude. This can be done only if dualism is maintained in a kind of half-life. Otherwise, the theses naturalism propounds would be unrelievedly physiological and would lack the wider interest they take on by claiming to reduce, while not quite eliminating, the "mental" to something neural. A similar tendency can be observed among those naturalists who talk about the "mind–brain."

The trouble is that if we close the gap between the brain and the perceived object in this way, it will be the effect produced in the "mind–brain" that is perceived, not the object outside our bodies. What is in fact proposed to us as such an imprint on the brain—formal topographic patterns, say, that correspond to the spatial relations to one another of the original stimulus-objects—is plainly not what we see. We are, in fact, quite unaware of these events in the brain and they lack, in any case, the kinds of qualitative differentiation that characterize the things that we do see. And even if this difficulty could be met, there would still be a question about who is going to view the effects produced in us by this process of transmission. The whole story evidently presupposes the existence of that homunculus who views the representations or images so produced. Plainly, though, this simply reproduces the puzzle that first motivated this line of thought. How, after all, would this homunculus be able to perceive the objects projected onto the inner screen? Presumably he would not do so through yet another process of transmission since that would entail the existence of another internal screen and that, in turn, would carry with it the same question. If we are consistent in our rejection of dualism and also recognize that an internal screen in the brain is a myth, this whole idea of perception as transmission also has to be called into question.

It hardly needs saying that this conclusion must be an extremely grave one, especially for anyone who acknowledges even minimally the one thing that we bring to these inquiries from our own pre-scientific experience: the fact that in perception something does *appear*. If the transmission theory of perception that is supposed to explain this fact cannot do so, then the theses of hard naturalism that deny that anything of the kind happens will have to be accepted. The only trouble, once again, will be that they are incredible and, in the bargain, cut the observational ground out from under any theory that asserts them.

There are a number of ways in which one might try to evade these conclusions. One such would be to ask whether what I see—my visual field—is not itself the final stage in the process that leads from the object to the eye and from the eye to the brain. There is, however, a very serious objection to which any such proposal is vulnerable. The things I see and otherwise perceive are mostly things that are outside my body. In what sense can such a perception be treated as something that occurs in my brain and thus inside my body? It has often been pointed out that if we try to go this route, everything we perceive—indeed, everything in the world, including our heads and our brains—will have to be assigned the status of a representation in our own heads. That would make the brain a very crowded place and it would certainly be much more like a Leibnizian monad than the brain of neurophysiology is supposed to be. To avoid such grotesque implications, we may have to agree that, of course, the things we perceive in the world—the world itself—cannot be in our heads; they remain where they are and it is only the perceiving that takes place in the brain. But in that case one would have to ask what kind of a relation on the part of something in the brain to something outside it is being postulated. This would have to be a relation that makes its external term visible and present to the person in whose brain the internal one is supposed to be located. The only thing that is clear about any such relation is that it is wholly unknown to neuroscience and to any other natural science on which brain research may draw.

Another way of trying to maneuver around these conclusions is exemplified by the kind of thinking about the mind and the body that goes by the name of functionalism. This is a position that draws on sources as various as Aristotle's *De Anima* and the distinction between hardware and software that comes out of computer science. It equates "mind" with what it calls the functional state of an organism at a given time—in other words, what it is up to in generating a certain output in response to a specified input. Because such functional states are commonly described in language that goes well beyond any purely physical idiom, a connotation of teleology may accrue to this account. By that I mean that the stark physicality of brain physiology is softened by making the brain the material bearer of all kinds of purposes that can sound very "mental." Nevertheless, it would be altogether misleading to encourage the idea that something other than familiar, garden-variety neural processes must be involved in this kind of functioning. On the naturalistic view, nothing else can be so involved; and naturalism could not claim that anything other than physical laws govern these processes without grave inconsistency with its own fundamental tenets. To this it must be added that if human beings are running on software of some kind as computers do, there would have to be some basis for the kind of claim that is regularly made in the case of purposive beings—the claim that what someone was *trying* to do failed for some reason and was not accomplished. It is hard to make sense of any such claim in the context of a fundamentally physicalistic conception of human being. Altogether, it would seem that, as far as any purposes are concerned that might be thought to be implicit in whatever function the brain performs, those would be indistinguishable, under the conditions that naturalism imposes, from the actual outcome of the relevant series of neural events.

It may seem as though hard naturalism could meet this objection very simply by pointing to what a person says or might say about what he is trying to do as the basis for the contrast that purposive action presupposes. That, indeed, is what it has done in an attempt to show how a transmission theory of human functioning can be plausibly construed in such a way that purposive effort can be distinguished from physical process as such. On such an account, what comes out at the other end is not anything mental or, indeed, anything that has to be described in the language of appearance or presence. It is verbal behavior. But that idea has already been shown to be a non-starter since speech as verbal behavior collapses into the very nature from which it has somehow to emerge if its goal is not simply to coincide with its actual ending. There is, accordingly, no need to examine it further here.

III

I have spoken of the official position of naturalism on the issue about brain function—a position that is undercut by the slide toward dualism that was implicit in the views just discussed. By contrast with the nostalgia such views display for a more familiar kind of contrast between the physical and the mental, the real thrust of a naturalistic theory of brain function is to dispense altogether with the idea of a human being as the being to whom something is present or given and to concentrate attention instead on the brain as a physical system. It may seem as though such a view would be unassailable within the narrow perimeter it stakes out as its proper domain. In fact, however, it can be shown to be inconsistent with the realities of neurophysiological inquiry. The paradox here is that this human being who has been bypassed continues to play an essential role in such inquiries. The reason for this is quite simple: the neuroscientist who conducts these inquiries is himself a human being and although, in one aspect, his work reduces all the people around him to the status of physical systems, it is not possible for him so to reduce himself or, as it turns out, his patient or experimental subject either. In fact, this unreduced humanity, understood as the capacity for having a world, comes into play at both terminals of this scientific work itself.

What has just been said suggests that there may be a difference between what the brain, at least as it is conceived by the natural sciences, can do and what we as human beings, in a condition of pre-scientific innocence, claim to be able to do. If there is such a difference, it might also suggest that the work of neuroscience is done on two very different conceptual levels—that of the brain as a distinct organ within a human being and that of a human being as a whole. The question would then arise of how we are to conceive the relation of parts to wholes when these wholes are human beings. In this section and the next chapter I hope to look further into this question. What needs to be established first is that there is this difference of levels in neuroscientific work. Beyond that, however, it needs to be shown that what we claim to be able to do as human beings forms an integral part of the very scientific procedures by which the scientific study of the brain is governed. If that is the case, then such capacities, which are so often dismissed as folk psychology, must surely command more respect in the eyes of the partisans of science than they might otherwise have been able to claim for themselves.

In order to understand this, one must bear in mind that the term "representation"—a borrowing from dualism—is much favored in contemporary neuroscience. Originally, a representation was something in the mind that was the object of an immediate awareness and that stood for something else that was not and could not be present or given itself. It was, accordingly, a synonym for other mentalistic words like "idea" or "thought" and even for "perception." As currently employed, however, a representation is simply the effect something in the world produces in the brain, without any requirement

WHAT DOES THE BRAIN DO?

that it have any properties in common with what it is said to represent. (It does suggest that the motive—doubtless unconscious—for using this mentalistic word in a physicalistic context may be to make it appear that the contrast between brain events and "mental" events is not quite as stark as it would otherwise have to be acknowledged to be.) In any case, when it is so conceived, a representation cannot be an object of awareness on the part of the person in whose brain it occurs, if only because patterns in the brain are not any more accessible to such a person than is the state of his liver. As a result, what is called a representation cannot really represent or stand for anything to that person; and it cannot, therefore, provide the basis for his beliefs about the world.

For whom, then, can it function as a representation? The only answer that can be given to this question is that it represents something to the neuroscientist. He does have a certain access, mediated by various forms of instrumentation, to the brain of our original perceiver as well as to the stimulus-object outside the latter's body. What actually happens is that in neuroscientific theory a subtle displacement of the notion of representation takes place in the course of which the neuroscientist is substituted for the perceiver as the being to whom such representations really represent something. Someone's brain state represents something to the neuroscientist insofar as it serves as the basis for inferences by him to other facts about the neural state or the behavior of the person in whose brain it is located. As things turn out, however, the role of this person is by no means limited to serving as the container for representations that do not represent anything to him. Instead, the neuroscientist proves to be dependent on this very person in a way that makes his collaboration a necessary condition for the pursuit of these inquiries.

How does this come about? It is a matter of the way the neurophysiologist identifies the brain state or event that is supposed to constitute a certain perception, say, or any other *ci-devant* "mental" state. Plainly, he can do so only by relying on what he is told by the person whose brain he is studying. In other words, when he isolates some state or event in the brain as the presumptive locus of a perception, he must address such questions as the following to his experimental subject. Does he perceive something and if so, what does he perceive when the brain event in question occurs? One thing is certain and that is that the answer the experimental subject gives—"I see a green wall," for example—will not be derived from an examination of his own brain by this subject since he would not be in a position to make any such observation even if he, too, were a neuroscientist. This report on which the neuroscientist has to rely is naive in the way all common-sense reports of perception are and simply states that some object outside the body of the person giving the report is a such-and-such. Evidently, then, this person is reporting a fact that not only is not a fact about his brain, but is also not a fact that the neurophysiologist who does have access to the brain of his subject is able to verify by means of anything that he can observe there. How can a naturalist explain this state of affairs? If perception is a physical event that takes place within the brain, it should, in principle, be as accessible to one person as it is to another. If it is not, how can it be an event in the brain?[3] The plain implication of all this is that in a vital respect the theory that identifies a perception with a brain state is parasitic on the very "naive" view it does not deign to discuss.

What is of capital importance in all this is the fact that this claim of access is not passed through the alembic of the physicalistic theory of perception either by the neuroscientist or by his philosophical co-adjutors. But since everything that he says about perception in his capacity as a neuroscientist applies to him as well, he is committed in principle to the thesis that his own perceptions of either that stimulus-object or the brain of his experimental subject are simply events in his own brain. But it is impossible to reconcile this thesis with the way the perceptions of the neuroscientist actually function in the context of his research. The events that take place in the brain may have been set in motion by light reflected from the object we are said to perceive; but those events themselves are, under the descriptions they bear in a neuroscientific account of them, utterly distinct from that object. By contrast, a perception as it is understood in the natural attitude is precisely the presence of that object to the person whose perception it is.

It begins to look as though this whole "scientific" line of thought simply assumes, just as common sense does and on the same pre-scientific evidence, that people generally, like this experimental subject, perceive objects and perceive them in the way that common sense assumes they do. This is the sense of "perceive" that entails that these objects and their properties are present to us as they are outside our bodies and our brains. If this were not being assumed by the naturalist, he would not be able to identify certain brain states as, in my example, perceptions of specific objects in the environing world. Those brain states themselves are at best causally related to the object that is perceived; under their scientific description, they have no capacity to maintain the kind of relation to that object in which it would be present to the brain. The brain, in fact, conceived simply as a physical system, cannot perceive anything. The idea that it can depends on the brain's being reincorporated into the context of the whole human being that does perceive objects. This would, of course, be the human being, not as reduced to a physical system, but as it is understood in the natural attitude. In that case, however, the edulcorated version of naturalism that permits the use of words like "consciousness" would have to be regarded as being a borrowing from the natural attitude and from the life of human beings who are capable of such perception. As such, it would amount to an abandonment of the central claim of hard naturalism about the brain and the causal character of our relation to the things around us in the world.

It seems likely that when the language of what were formerly held to be mental states is applied to the brain, this is done on the strength of a prior conviction that the brain must be what performs the functions traditionally assigned to the mind. The result is to make it appear that there is a smooth continuity between brain function as neuroscience describes it and the kinds of human functions that have traditionally been called "mental." On the strength of this transfer of the functions of the whole human being to a part thereof, the neuroscientist can provide the processes he studies in the brain with the whole rich context of perception as it is ordinarily understood. He then describes those physico-chemical processes in language that presupposes all those understandings. In this way, we are given the assurance that a full account of perception has been provided without having to explain just how the processes that neuroscience studies issue in something that has any claim to be called a perception. Otherwise, if the neuroscientific account confined itself to the language of physics and chemistry in its descriptions of brain processes, the implication would be all too plain that no one, including the experimental subject, ever sees or perceives anything in any sense of "perceive" with which they are already familiar. Among other things, this would make it impossible to distinguish between a person who is blind and one who is sighted or at any rate to do so in a way that is not visibly parasitic on a prior (and very different) understanding of that distinction.

IV

Earlier in this chapter, the kind of dilemma that confronts the scientific study of the "mind–brain" was set forth. What it comes to is that there are two main philosophical interpretations of what the brain does, one of them dualistic and the other naturalistic. If the former is adopted, it brings with it all the paradoxes that were set forth in Chapter 2 and, at the same time, reproduces at one remove all the puzzles that inspired these inquiries to begin with. But if, instead, the naturalistic interpretation is favored, there will be a simple denial that there is anything to be explained other than the physical processes that occur in the brain and the central nervous system. This means that naturalism must require of its adherents disclaimers of any familiarity on their part with the way things look and feel—disclaimers that, as noted, put a strain on credulity. It also makes its own status as an account of processes in the brain (or of anything else) utterly mysterious since they have been detached from anything that could be called observation. One does not have to be an empiricist to find it hard to believe that any knowledge, scientific or otherwise, can be gained in the absence of any observational access to the things this knowledge is about. I have argued that this strange situation can be accounted for only on the assumption that the naturalist is effectively doing business in the natural attitude without, of course, ever admitting that this is so.

It is hardly surprising that, when faced with this kind of dilemma, many partisans of naturalism cling to the hope that the science of the future will resolve all these puzzles. This can only be an act of pure faith since no reasons are given for thinking that the kind of difficulty that has been encountered can be resolved by the kind of inquiry in which the natural sciences engage. There have also been suggestions that there must be something in the way we ourselves are constituted that makes it impossible for us to understand how the brain can produce the kind of conscious life or experience with which we are familiar.[4] Strangely, however, this obstacle is apparently not so serious that it prevents us from understanding that in some way or other the brain must hold the answers to all our questions. Neither of these responses to the situation in which we find ourselves changes it in any way or offers any reason to think that it will change in any significant respect. Nor do they explore the possibility that the trouble may lie in the way our questions themselves are conceived and in the philosophical presuppositions that inform them.

What I want to do in this section is to offer a number of suggestions about what has gone wrong in the kind of thinking that leads to this impasse. The deepest source of the trouble seems to lie in an obstinate insistence that the methods of the natural sciences on which naturalism has staked everything are of universal application. They simply must be right and so their claims must be driven through, whatever resistance certain contexts of inquiry may seem to offer. The alternative would be to conclude that there are some things about ourselves as human beings that we simply have to accept without being able to reconstruct them in the idiom of the natural sciences.

What I especially have in mind here is the idea that seems to preside over so much scientific work that is now being done. This is the idea that it ought to be possible for the natural sciences, after bringing so much of the world under their characteristic modes of conceptualization, to come around again, on their way through the world, to the place from which they set out. This is the place that is variously described as that of consciousness or experience or mind, but I prefer to call it the locus of presence. In a way, of course, we have never left it since scientific inquiry is itself a conspicuous expansion of the zone of openness we inhabit. Nevertheless, since nothing like that is acknowledged in the account natural science gives of what it is about, it is as though it had been left behind. The paradox is that science claims to be able, not only to come back to the place where perception and thought come into being, but also to see it again for the first time. That means that it proposes to reconstruct out of its own conceptual resources the situation out of which it has itself come. It will, of course, have to make some drastic changes in the architectural design of its birthplace; and if a commemorative plaque recalls the quaint structures, with odd names like the ones just cited, that once stood there, that is probably as much as one can hope for from the sponsors of this line of thought.

What I am proposing is that the conspicuous failure of naturalism to achieve this goal (and especially the failure of the concept of transmission as an explanation of the fact that things show themselves) justify us in concluding that presence has to be treated as a primitive concept for the purposes of such inquiries as these. What this means, in the first instance, is that we must give up the idea that presence itself can be explained by the application to it of the explanatory methods of the natural sciences and thereby collapsed into the object domain of those sciences. To claim that it can be is to involve oneself in a quite remarkable contradiction. It is, first of all, to ignore the fact that this very claim, like the inquiry of which it forms a part, is made from a position vis-à-vis what it is about that can only be characterized by appealing, explicitly or implicitly, to the fact of presence. That fact, as the opening-up of the kind of space in which a thing can show itself as what it is, cannot consistently be repudiated by anyone engaged in the inquiries for which it serves as a necessary condition. Any attempt to negate presence by reducing it to what is present can only be described as an attempt to reverse the "event," if there ever has been something of that kind, in which nature becomes a world. Such an attempt is destined to fail.

To argue that presence is prior to all scientific forms of inquiry is not to deny that in many ways all sorts of empirical facts about us—the way our eyes are positioned and the way they move, for example—will make a difference to the way things are present to us. The point is rather that none of these facts can account for presence itself although once it is realized, they figure in it in a variety of ways. Another way of putting this would be to say that the age-old naturalistic assumption that Marx expressed—by saying that matter comes first in cosmic evolution and that it brings into being consciousness or mind—has to be given up. This is not because consciousness came first in the way that matter is supposed to have come first. It is because there is no way of showing that presence *can* emerge out of the processes with which physics, as the foundational natural science, deals. The cosmological status that should be accorded to presence itself is a question to which there is no obvious answer; but some relevant observations will be offered in Chapter 6.

What can be said is that presence constitutes the distinctive ontological condition of those entities—notably, but perhaps not exclusively, human beings—that we have hitherto conceived in terms of such notions as "mind," "consciousness," and "experience." In other words, presence, once it is in place, sets the character of the entities for which it makes having a world possible, and it does so in a radical and pervasive way. This means that the functioning of those entities and of the organs and parts that make them up are, in one way or another, all decisively shaped by this way of being in the world and of being in it with other entities. The eye of a living creature, for example, is inconceivable otherwise than in a milieu of presence and we may suspect that the same is true of the brain. All inquiry effectively begins in that milieu, whether that fact is acknowledged or not, and it proceeds on the tacit assumption that we are beings for whom the entities that make up the world are there. The paradox that is the occasion for this book is the fact that the methods of inquiry that have been developed for dealing with natural phenomena, when generalized, make no place for the presence on which they are themselves founded. When these same methods are applied to the human beings that pursue such inquiries, the result is the extreme confusion and incoherence that naturalism has produced in the intellectual world generally.

Any such thesis as this gives rise to so many questions and challenges that one scarcely knows where to begin one's defense of it. Perhaps certain prevalent misunderstandings of the whole notion of a thing's being present offer as good a point of departure as any. The misunderstanding to which I would especially draw attention has to do with the distinction between a thing and its character as a phenomenon—something that shows itself and thus is present. One reason for this may be that we still think about all these matters in a reifying idiom that seeks to explain everything by postulating some sort of "device" or "mechanism" that will produce the phenomenon that puzzles us.[5] We seem not to be able to give due weight to the fact that things that are present to us do not emerge as such from some kind of processor in our heads. They are there for us at a certain distance and in a certain perspective and orientation.

This last point harks back to an observation that has already been made. That is that it is not just by virtue of being there for us that the things in question depart from the model of objectivity. Objects as they are conceived by the natural sciences are not supposed to have any oriented character. They do not face one way rather than another and there is no difference between the outside and the inside of objects so conceived. And yet this oriented character is precisely what constitutes a thing as a phenomenon and as the worldly vis-à-vis of a perceiver. This is, moreover, the distinction that really counts when it comes to drawing a line that separates the phenomenal domain from the world of nature. Unfortunately, it has been distorted out of all recognition by being understood as a contrast between what is in the mind and what is outside it.

Instead of exploring further this remarkable disaffinity between things in nature and in the world, the line we take typically treats the unique disclosure that admits us to a domain in which we deal with things as the kinds of things they show themselves to be as being of negligible importance by comparison with some other way by which we come by the knowledge we have of the world. Although a priori reasoning which was once thought to be such a way has long been discredited, "theory" now seems to have much the same attraction, especially when it places itself under the auspices of "Science." Even so, it must be plain that whatever objects we have factual grounds for postulating can be traced back to those that perception affords us. Unfortunately, instead of trying to find an appropriate mode of expression for this radical fact of presence as the disclosure of an object domain, we prefer to try to account for our knowledge by moving these objects back and forth on a trajectory that runs from where they ostensibly are in the world to some putative locus—

mind or brain—inside us. There they become representations of themselves—pictures inside the mental gallery of what is supposed to exist outside it. But of course this "outside" would be wholly mythical if it were not for a lingering understanding we have of its status before it was moved indoors and turned into a picture or "representation" of itself.

What has just been said points to another thesis that is bound up with the one that makes presence a primitive concept for the purposes of understanding brain function. That thesis is that there is really only one set of objects with which we deal and they are the ones that are given to us in perception. To demote them to some purely subjective status or to disregard them altogether for purposes of theory-construction is inevitably to make our commerce with objects, generally, incomprehensible across the board. There simply is no other source from which our everyday familiarity with, or our scientific knowledge of, the world as an object domain could derive. This is not to deny that we may perform all sorts of conceptual operations on these objects by which they are denied some of their properties—color, for example—that may be irrelevant for specific explanatory purposes. We also postulate that material objects have parts that are too small to be perceptible and that the molar properties of such objects can be explained by the behavior of these parts which are accessible only through elaborate systems of instrumentation. But however elaborate these operations may be, they always presuppose the reality of the things from which we all set out and which most of us never leave. Among all these objects and their constituent parts, including the ones postulated in theory, there is an empirically determinable continuity in space and time and in causal relatedness. This means that all the contrasts we have occasion to make, including those between what is real and unreal and between what is "inside" and what is "outside" have to be made *within* the domain of what is present to us in perception if they are to be intelligible. There simply is no contrast that can be made between the entities which, according to the transmission theory, are at best imperfect mental simulacra of objects in the world and those objects themselves. When we want to determine what is really the case in the world, the things that count are the things that we can perceive. And even though perception may indeed be misleading at times, it is only through a corrected perception of the same objects that it can be proved wrong.[6]

The great fact that makes an acceptance of anything like the primitive status I am according presence so difficult is not just that we cannot bring the incidence of presence under any scientific laws; it is also the fact that the way our bodies and our brains function in what I have called a milieu of presence lies, in good part, outside any nomological explanation that is available to us. That fact makes it almost certain that the idea I am proposing will be dismissed as altogether too fanciful to be taken seriously. Although I am not presenting this idea as one that I expect the natural sciences of the future to be able to explain to us, those who are most likely to react in this way typically *are* enamored of futuristic science. It seems fair, therefore, although admittedly in an *ad hominem* sort of way, to ask them how much more fanciful my proposal really is than the sort of thing many philosophers routinely imagine as being possible for some science of the future. In a famous example, they imagine a brain in a vat into which stimuli are fed that create a totally false environing world for that brain. Surely, if it is held to be possible in principle to give a brain access to a coherent world that just happens to be completely false, one must ask why it should not be equally possible for the brain or for the human being with a brain to have such access already to a world that really exists?

It will, of course, be objected that, in the case of the brain in a vat, what we bring into being is really just a phantasm in the brain whereas in the other case it would be something actual in the world. The difficulties about explaining how such "experiences" could be *in* the brain are already familiar. This objection also conspicuously misses the point that in the latter case, instead of bringing into being a second set of objects that are then declared to be in the mind, it would only be the presence of actual objects in the world that would need to be brought about. To be sure, if we want to say that the brain does this, we would have to conclude that it must be functioning in ways that are not only unknown to the natural sciences but are very likely to remain so. But this is the same assumption that we make when we claim that our experience of the world is produced by brain processes since no one is in a position to cite the law of physics or chemistry or biology in accordance with which this is supposed to be done. Many people would probably want to insist that this inability will eventually be overcome by the progress of science.

This is the deep paradox by which naturalistic theory is beset in this area. It has to choose between a dualistic "mind–brain" contrast, that at least seems to give substance to its claim to explain presence, but does so by giving it a new object—a representation—and a no-nonsense physicalism that has a kamikaze quality about it. Since neither of these options is at all attractive, we have to wonder whether there has not been some vice in the argumentation that issues in this set of alternatives. The suggestion I want to make is that there is indeed such a defect and that it consists in the fact that neither dualism nor hard naturalism gives enough attention to the role of truth as a characteristic of that which is supposed to be produced by the functioning of the brain. Whatever that may be and by whatever name it is designated, it is supposed to be capable of being true. That means that it is not simply a thing or a state of affairs, whether in the brain or in the mind; it carries an essential reference to something in the world in relation to which it can be found to be true or false.

The trouble is that both these theories fail to make any satisfactory provision for the kind of access to the world that such a relation presupposes. In the case of dualism, this failure is due to the mediated character of the mind's relation to the world; its only objects are its own representations and what these are supposed to represent remains forever out of reach. Naturalism fails in another way. Dealing as it does in brain states which stand in causal relations to things outside the brain, it is in no position to postulate any relation of representation other than the one that is mediated by the scientific theory of brain function that the neuroscientist brings to bear on someone else's brain function. And even this, as has already been shown, presupposes a quite different relation in which the neuroscientist stands as a human being to the relevant things in the world and to his human subject.

It is in fact only in the natural attitude that the requisite openness to the world that makes truth possible can be found. It is not surprising, therefore, that it is covertly resorted to by the partisans of both dualism and naturalism who thereby set aside the crippling limitations under which they would otherwise have to work. There is something almost comical in the way implicitly dualistic idioms reify "the mind" as though it were simply another organ of a rather special kind that can on occasion play tricks on "us"—read "me"—and has to be called to order by an "I" that is evidently not identical with it. This is the same duality that surfaces in naive talk of "introspection" as "looking into the mind"—talk that unavoidably presupposes some sort of contrast between what so looks—from where?—and what is looked into. The persistent violations by naturalism of its own self-imposed restrictions on acknowledging anything that involves presence or givenness have been cited too often to make it necessary to review them again. The plain fact is that both these positions live off resources that are not supposed to be available to them. By doing so they tacitly acknowledge the natural attitude they have repudiated as the background against which they do business.

An appendix on pain

I can imagine someone—a philosopher—who has read this far and whose reaction might be expressed in something like the following way. "These lofty pronouncements are all very well," he might say, "but what exactly do you propose to do about pain?" Pain has, indeed, long been the King Charles's head of philosophers of mind; it stubbornly asserts its reality against all the reductive strategies that have been used to domesticate it within some large philosophical synthesis or other. It has been variously described as "free-floating" and "anomalous"; and it seems to be at home only in some setting that welcomes the most heterogeneous *qualia* and allows them to dangle promiscuously off the world of law-abiding properties. That setting can hardly be one that naturalism would find congenial and so the effort that naturalists have made to identify pain with the activation of c-fibers is at least understandable as expressing sheer exasperation at a habitual offender against the norms of scientific rationality.

It may seem a little incongruous after all the criticisms that have been directed at naturalism in this study to offer it assistance against a stock argument that dualists use against it. Nevertheless, there is a way of accommodating pain within the position I have been sketching; and I think it should be more attractive even to naturalists than an outright regression to dualism. An example may help to introduce the idea I have in mind. Suppose that I carelessly step on someone's toe and that person gives a yelp of pain. The contact of my foot with that of this person is certainly in some sense a physical event and so are the neural and muscular events that follow upon it. All these events are also what we call "public": they occur in a domain to which everyone has access. The pain that our man suffers, however, is not accessible to everyone; only he feels it and so it is in some sense private. Normally, we would say that this pain is caused by the pressure of my foot on the victim's foot; but if this pain is not discoverable by others in the foot in which, nevertheless, it is felt by him, there seems to be something strange about this kind of causation. It is as though we were dealing with something that has gone off the maps we are using but which we are unable simply to deny as though it were some sort of illusion. Pain simply cannot be generically referred to the imagination in the way physicians do when patients describe pains they cannot account for in any familiar way.

This is where the suggestion I want to make may be in order. Is it not possible that the nature of pain itself is being misconceived in a way that at least partly accounts for the anomalous status it seems to have in examples of this kind? We tend to think of pain as a funny kind of quality—what has been called a "nomological dangler"—because it has no place in the kind of causal explanations that the sciences offer. But instead of trying to treat pain, in spite of these peculiarities, as the effect produced by the pressure of my foot on that of this other chap, we might well regard it as being the modality in which that event and the parts of his body that are involved in it are present to him. In that case, the "physical" element in this incident—the contact of my foot with his—would still be a necessary condition for the occurrence of the pain and it would still be the cause of the pressure on his foot. The pain itself, however, would belong in another context—that of the life of another being that has not only a foot but a body (and a world) and registers what is happening to it in a variety of modes, one of which is pain. After all, when we are in pain, something hurts—our head, our stomach, our toe—and even if this rough localization of the pain may not suffice for purposes of medical diagnosis, it is enough to give that pain a place in the world which it would not have if it were really just "free-floating."

There is a way in which anyone who happens to step on someone's toe can reconcile himself to this account even if he has a strong pre-disposition in favor of hard naturalism. Such a person may well *see* what is happening—his foot coming down on that of the other person. If he does this deliberately, he may well enjoy the sight of the victim's discomfort. If he does so unintentionally, his perception may be tinged with guilt. In any case, his apprehension of the fact in question need not be neutral even if it is not as anguished as that of his victim. It might also occur to him that just as this is the way he comes to "know" what has happened, so the pain would be the way his victim does, especially if he does not otherwise perceive this event. What I am getting at here is the fact that these two people in their different ways register what happens when one of them steps on the other's foot. Pain is a way in which a part of our own body can be present to us; and in this sense there is an affinity between vision and pain as two ways of having a world. In neither case is the causal sequence between the physical and the non-physical parts of this event—the pressure and the seeing or the feeling of pain—as straightforward as we may think it should be. But for just this reason there is no basis for the supposition that pain is especially anomalous and in a way that requires that it be sequestered in some special place like the mind.

6

HUMAN BEING AS THE PLACE OF TRUTH

I

The criticisms I have made of the views discussed in earlier chapters imply, not only that naturalism severely misrepresents important aspects of human nature, but also that there must be another, better understanding of what we are as human beings. Some of the lines of such an alternative view have already been sketched, but in this chapter I want to see if I can pull them together to form a somewhat fuller picture of the matters that I claim have been so seriously misconstrued. Although nothing like a full defense of this conception can be attempted here, it should at least be possible to give a reasonably accurate idea of the kind of concept that would be congruent with the conclusions reached in preceding chapters. At the same time, I hope to correct certain misunderstandings that could result from some of the formulations I have used up to this point.

The account of our human mode of being that is being proposed here replaces the contrast between mind and body with another contrast that is no less fundamental but very different in character. It is the contrast between entities as present to someone (and, in principle, to anyone) and these same entities as not present to anyone. My claim has been that what we have thought of under the rubrics of "mind" and "the mental" comes down to the first term in this contrast: entities as present and this means as present to another entity (and in certain cases to themselves). The most obvious point of difference between these two contrasts is that the second does not try to deal with the puzzles connected with our way of being in the world by postulating some additional entity or some other place in which the relevant functions are supposed to occur. Adding another story—the mind—to our familiar, corporeal domicile, as dualism does, just reproduces the same old difficulties. This account does not, however, simply assimilate our mode of being to that of the other entities we encounter in the world so that the only remaining questions about our nature would be scientific ones.

What was said in Chapter 3 about our dealing with things in the world as what they *are* can help us to understand this second contrast more fully. The point made there was that the things around us are not just other items in the totality that makes up the world and in which we are included as well. Nor do they simply act upon us in the various ways for which their causal powers suit them. They are also there for us and are dealt with by us in terms of the characters that make them the kinds of entities they are. For this to be possible, they have to be available to us in those characters so that we can recognize and identify them as what they are in perception and order them in memory and in thought. But because all this is so familiar to us and because the things around us are the same ones that the sciences deal with in their own characteristic ways, it is extremely difficult for us to take seriously the idea that there is this difference between a world in which there is no presence and one in which there is. Instead, we do our best to identify the latter with the former. In fact, however, those sciences make no place for the world as a place of openness or a milieu of presence; and as a result we are in the strange position of living in a world that we think of in terms of a model in which *we* have no place. It might be said of our situation as it emerges from the preceding discussions that everything in it remains the same and yet everything is different. Everything remains the same because the entities with which we have to deal are the same ones we were familiar with under the prior dispensation. Everything is different because we are now in a position to recognize the distinctiveness of the mode of being—ours—in which these same things can be there for us as the kinds of things they are.

There is an even deeper paradox here. The natural sciences that propose the model that turns out not to make any place for us really are just as dependent on the one that does as are the supposedly more quintessentially human domains of thought like ethics and the arts. Scientists are, after all, human beings even though they often seem to want the waters of theory to close over them in that capacity. The point here is simply that the sciences rest on observation and thus on presence both for the substance of their theorizing and for the tests to which it has to be put. It is true that there are those *esprits forts* who are undeterred by any such argument. They are quite prepared to reduce perception and thus observation to a coded stream of sensory "information" and scientific theory itself to the words and symbols that emerge at the other end of what might be called the epistemic canal. On such a view, theory formation would be quite literally a neural process that proceeds in the third person without there being a need for anything actually to appear. What is thus envisaged is, in effect, a version of Searle's Chinese Room paradox understood as the way our own brains and bodies work—that is, in complete independence of any semantic element in our interaction with the world around us.[1]

I have spoken of a contrast between a world that is a milieu of presence and one that is not; but it would be better to reserve the word "world" for the first term in this contrast and to use "nature" for the second. In any case, it is clear that this is a contrast that can be understood only by someone who is in the world as a milieu of presence. Such a being can imagine a nature in which nothing is present to anything else—the kind that existed during most of the immensely long period of cosmic evolution that, on present scientific accounts, preceded the emergence of life and "consciousness" in relatively recent times. In a way, such imaginings could be said to be an intrusion on the part of the being who so imagines the past into a domain that is otherwise without presence. That being is, of course, not really there; and this is, in any case, the only way in which we can give meaning to the contrast with which this chapter began.

The question all this leaves us with is still what kind of entity a human being is. Dualism and hard naturalism offered characterizations of that entity, but they have been found to be deeply unsatisfactory. But if we are not prepared to endorse a conception of the world without any entity in it that functions as a subject, we have to begin working out at least the main lines of the concept of such an entity. This means that the peculiar way in which our "mental" functions and most notably that of perception have been disjoined from the things in the world that are supposed to be their objects must be set aside. What this in turn means is that we ourselves will have to be conceived in some way that does not simply shut us up in ourselves as happens when lines of thought like those just described are in control. Our being undoubtedly has boundaries, but they are not the epidermis, as some now seem to think, nor are they the "walls of the mind."[2] What is urgently required is an interpretation of the ontological status of a human being—the kind of entity it is—that does not issue in either of these kinds of encapsulation of our being.

In this chapter there will be a modification of the terminology for referring to the characteristic ontological structure of human beings that has been used up to this point. The term "presence" will be retained, but it will be reserved for expressing the status of those entities that are present to someone. "Transcendence" will apply to the entities—human beings—to which other entities are present.[3] In this way the asymmetrical character of presence will be recognized. This usage, with its links to the verb "to transcend," should not be interpreted as imputing some special kind of agency to human being by virtue of which it may be said to transcend itself. Instead, it is transcendence that makes agency of whatever kind possible. Accordingly, the use made here of the notion of transcendence, both as a substantive and a verb, will be entirely neutral with respect to all further questions that can be raised about what transcendence and/or presence are and how they come into being.

II

The great difficulty that awaits those who attempt to work out such a concept as this is the way the concepts of human being that we already have insinuate themselves into the new ones. The former are mostly dualistic in character and, as such, products of the splitting of human being into two distinct entities—the body and the mind. This is to be replaced by a conception of human beings as unitary entities that are "in the world" with other entities in the special sense of this expression that turns on the notion of disclosure. We are thereby committed to the idea of ourselves as entities that stand in a distinctive relation to others; and although this seems very straightforward, such a use of the concept of relation is really a good deal more problematic than it may at first appear to be. I want to take up, first, the question of whether being-in-the-world can satisfy the general criteria by which the concept of a relation has traditionally been defined. Then I will turn to the more specific issues posed by the understanding we have of perception as a relation in the terms set by what I have called the transmission theory.

Under the first heading, the main issue that has to be resolved is whether presence can properly be said to be a relation at all. One doubt about this is suggested by the fact that presence does not seem just to connect two terms with one another as a relation is supposed to do. It is not, for example, a side-by-side relation like that of cars that are parked next to one another. But then what kind of relation is it? We want to say that it is a relation in which one thing or entity is present to another. Since these entities would presumably be located somewhere and would therefore already stand in at least a spatial relation to one another, it sounds as though presence would *add* something to that relation by making at least one of these terms present to the other. Even if we accept that this "something" is not to be conceived as itself an entity or a property in any familiar sense, this may well be thought to be more than should be expected of a relation, at least as that notion has traditionally been understood. Clearly, we need to take a closer look at the concept of a relation itself if we are to determine how serious this kind of objection really is.

The concept of a relation forms part of a table of categories that has played a great role in both ancient and modern philosophy. A table of categories is a taxonomy of the kinds of predicates that can follow the verb "to be"; and the philosophical tradition has made the concept of substance or thing the category that serves as the foundation for everything else that something can be said to be. But since there are many substances, a concept is needed that designates the ways these substances are with one another in the world. They can be next to one another in space or one can come after another in time; and both Aristotle and Kant provided categories to cover facts of this order. The presumption has been that the category of relation as so conceived could also serve satisfactorily to deal with all the kinds of facts that turn up in connection with human beings. On this view, the human body does not stand in any relation that is inapplicable to other bodies simply as physical objects; and the presumption has been that the same must hold for the human "mind" as a substance of a special kind. Certainly, no categories were provided that were specifically designed to apply to mental substances rather than to material ones.

Nevertheless, there are some fairly well-known difficulties about the use of the category of relation in connection with "mental" functions. One of these is the possibility of error—that is, of our being mistaken in the beliefs that ostensibly set up a relation to something else in the world. When this happens, and the worldly term whose existence we affirm turns out not to exist, the "relation" we so postulate will not be a relation *to* anything and thus hardly a relation at all. Far more serious for present purposes is the fact that the terms of a relation are supposed to be such that they would still exist even if any given relation in which they stand to something else were to lapse. This requirement flows from the idea that a relation holds between two or more things or substances and that these do not depend on their relations to other entities for their own existence. But, on the view proposed in this book, the concept of mind is replaced by that of the presence of other entities to a human being. This conception is fundamentally different from Descartes' idea of the mind—"the thing that thinks"—as a mental substance and thus as something that needs no other entity in order to exist itself. If presence is a relation that other entities stand in to an entity like one of us, however, "we" (or rather, in each case, "I") as a term in that relation would not be prior to the relation itself. There would, in short, be no entity like what we normally describe as the "mind" of the person in question that would exist independently of the presence to it of things in the world. In that sense, the usual concept of a relation would be subverted.

A somewhat similar line of reasoning can be developed with respect to the other term in what is now beginning to look less and less like a relation in any familiar sense. Although an object like a house certainly does not borrow its reality from something else in the manner just described, it, too, would owe its character as a term in a perceptual relation to a human being to the fact of presence. This fact may be missed when we think of relations, as we tend to do, by using examples like a book on a table or the earth and the moon—examples in which we ourselves are not directly involved. It is the two things themselves in their massy materiality that are the terms of such a relation which might even receive the ultimate compliment of being called a physical fact. But if the relation is one in which *I* perceive something—a book or a planet, say, or in this case a house—the physicality of this relation will be more than a little dubious. In such a case, the house will figure in the relation as something that shows itself—in this case, to me—and showing itself simply is not a physical fact even though what does so may be and typically is a physical object.

It should now be clear why the concept of relation has a problematic character as a rendering of the concept of presence. This is because it presupposes that there is a term in that relation to which something else—the other term—becomes present. The inevitable question must be: what is that first term? The classical answer has been that it is a soul or mind and with that answer we are once again working in the terms set by the thing-ontology although in this case the "thing" is, as Heidegger put it, a "spiritual" one. When we do that, the fact that this relation is typically an asymmetrical one will also strongly encourage a conception of it as transitive in the sense that presence somehow passes from the mind to the "object" rather like the beam of a flashlight. In fact, there seems to be no basis for such a conception of the genesis of presence. To the degree that one can conceive it as a relation between two terms, there is no natural property of either the one that is present or the one to which it is present that can account for the former's becoming present to the latter. And yet the one *is* present to the other.

All this suggests that if presence is a relation, it is a highly distinctive and in some ways anomalous one. It does not simply connect two fully formed objects; it supervenes on the one that is present and it in some sense constitutes the other—the one to which what is present is present—as the kind of entity that we think of a human being as being. Once again, this is more than a "relation" is expected to do; and if a relation holds *between* two terms, then this would be a decidedly unusual kind of "between." One way of dealing with this apparent anomaly would be to conclude that this relation is an internal relation and, as such, at least partly constitutive of its terms. That idea will form part of the conception I will outline in the next section, but because presence is more than and different from a logical relation, a rather different line of approach seems to be called for.

If we now bring these considerations to bear on the perceptual relation that is supposed to hold between something in the world and a "mind–brain"— the relation that has been understood as a transmission—some interesting conclusions emerge. If this relation is thought of as obtaining between a representation, however conceived, and a thing in the world, and if this relation is assumed to conform to the requirements implicit in the category of relation, there would be absolutely no reason to think that the former would be a disclosure of the latter. Each of these terms would have its own characteristics in complete independence of the other; and this would be true even if the representation were a mental image or picture of something and that something were exactly similar to what is in the picture. That conclusion would follow even more obviously if the representation were a brain state of some kind with no resemblance at all to the thing it is supposed to represent. As was explained in Chapter 5, the only sense in which this relation would be representative would be by virtue of the causal connection that a neuroscientist may be in a position to establish between these two terms. Even then, both the terms of the relation would have had to be disclosed to him already and independently of one another.

The conclusion for which I am arguing here is not that disclosure, understood as something's being there for something/someone else, has no relational character. It is that in that relation there are indeed two terms but one of them—the being for whom the other is disclosed—is not separate and distinct from the latter in the way required for the category of relation and presupposed in the idea of transmission. In this aspect, it is simply the being-there of that thing; and in the case of a human being that perceives or otherwise discloses such a thing, the being-there of that thing becomes constitutive of the kind of entity it is. This is an internal relation, but in an ontological rather than a purely logical sense. Another way of putting this would be to say that the facts that have been thought to justify the introduction of a concept of mind need to be conceived in a very different way. Instead of being treated as the internal states of a special kind of entity—the mind—they need to be seen as a new kind of relation among the entities that make up what we call the world. If we are to think of this kind of relatedness as emerging in the course of cosmic history, then it is an "event" that has gone completely unnoticed in the standard scientific accounts we have of the processes that make up that history. It is a relation in which one entity is present to another and sometimes reciprocally so. The emergence of such a relation would be at once the genesis of the world understood as what we are in and of the kind of entity that can be in the world in the mode of disclosure—that is, in the distinctive pairing with things and people in the world that has been described in earlier chapters.

These assertions will seem very bold to many philosophers because they have to do with our great vis-à-vis—"the world"—and not just with what is going on at our end of our putative relation to it. To claim, as I have done, that the world is somehow intimately bound up with human selfhood makes things even worse and will undoubtedly arouse fears that any such idea would foredoom the possibility of distinguishing between what is really "out there" and what is "in us." Actually, all that has happened is that I have tried to put into words something we are all so familiar with that it has seemed unnecessary to bother ourselves with such a task. It seems, however, that when matters that have been understood only tacitly are put into words, they often sound strange and exotic; and we have trouble accepting the idea that what they express is what we have been so familiar with all along. One way of dealing with the sense of strangeness we feel when the "worldliness" of selfhood is asserted is to complain that some logical outrage is being committed. Nothing that has been proposed here, however, is really any stranger than the dualistic picture we typically invoke when we ourselves have to give an account of what is going on in our "experience." And when we get past this first baffled reaction to these ideas, we will find that they are actually a lot closer to what we call common sense than it seemed at first possible that they could be.

What is being asserted is simply that being-in-the-world has to be understood as the genesis of objecthood.[4] "Objecthood" is not to be confused with objectivity and it is best understood through the etymology of the Latin word—*objectum*—from which both these words derive. "*Objectum*" means literally something that is thrown across one's path; it is in one's way and confronts one. In ancient Latin, the word also took on a wider connotation in which it expressed the visibility and thus the presence of whatever may be in one's way in this sense. This is also what Heidegger tried to express by the way he uses the word "world"; and he even invented an expression—*Es weltet* (it "worlds")—that was designed to capture, by the use of an impersonal verb, the utterly familiar fact of something's being there.[5] Because we are so familiar with things as *objecta* in this sense, it is hard for us to understand that this is a condition that accrues to them in their relation to beings that can perceive. It is also in that relation that things can be found to exist and to be whatever they are; and this is the reason why their being is so closely bound up with presence. One can, accordingly, say that presence, so far from being sequestered in the inner recesses of the mind as consciousness and representations have been supposed to be, functions more like an extra dimension that lets things be set over against us—in other words, as objects or *Gegenstände*. The upshot of the matter is that our being has to be understood in terms of an absolutely distinctive pairing of self with world in which each term owes its character in large measure to the other. This is a far cry from all Cartesian conceptions of the mind as a mental substance that just happens to be in the world and might equally well not be.

III

It has become evident that the whole idea of applying the transmission metaphor to perception was deeply wrongheaded and that a search for alternatives to it is more than justified. More specifically, the preceding discussion suggests that a satisfactory alternative would have as much to do with the world as it would with anything that we can plausibly claim to generate on our own. There are hardly any precedents other than Heidegger for such a line of exploration, but there is one that has, at the very least, permanently influenced the language in which we describe both cognition and its object and especially the Between that seems to be so central to the idea of presence. I have in mind the attempt to understand all of these matters in terms of light.[6] Light is a necessary condition for the visibility of the things we see; and it is therefore well suited to express, metaphorically, the character of presence itself. Because this conception of light developed long before its physical character was understood, there is every reason to think that it had been principally a concept of light as illumination and of the resulting visibility of the objects so illumined. Indeed, if it were not for the fact that illumination is confined to things that we see, it would be tempting to identify it with presence. But because presence is not so confined and has modalities in which it is linked with hearing and with touch, not to speak of presence-in-absence, the metaphor of illumination can hardly do justice to it. Nevertheless, since vision is the principal sense modality in which things are present to us, light as a necessary condition for their visibility has an understandably privileged role in our thinking about presence generally.

In the modern period, this conception of the ontological function of light came to be largely confined to the mind where it figured as the transparency to itself of the mind/soul. This meant that everything in the mind was necessarily manifest to the mind itself and nothing could be hidden from it; this transparency was attributed to consciousness as "the light of the soul." The downside to this use of the metaphor of light is that it commits one to the very kind of mentalism that has already been shown to have such a detrimental effect on thinking about these matters. In this picture, the world as such is not understood as a zone of openness in the way the mind is; and yet that is just what the line of thought developed in this and earlier chapters appears to require. What seems much more promising would be a return to the notion of the world itself as a milieu of presence that a human being is in. This is the conception that preceded that of an intramental consciousness that effects the mind's transparency to itself. Of special interest in this connection is the fact that, in spite of the connotations of mystery and mysticism that still cling to this whole set of ideas, it also has clear affinities with the natural attitude and thus with the sense we have of being in the world and in the presence of the things that make it up. That the idea of light still has an important metaphorical role to play in such a revision is evident in the language of Heidegger's own rendering of the natural attitude. He speaks not of consciousness, but of a "clearing" within the world; and in the word he uses, *Lichtung*, the *Licht* is, of course, "light."

The philosophical core of an ontology of light is the idea of being-in-the-world as involving a special kind of Between that makes possible the presence of one entity to another. In such a Between, the character of the entity for which things in the world are there would be intimately bound up with the character of that world itself. There is an ambiguity that attaches to this idea of understanding human being out of its world. If it were taken to mean that human being is to be assimilated to the mode of being of the entities that are within the world, it would be profoundly misconceived. What is intended is the very different idea that our mode of being is constituted by the presence to us of things within the world and that its own character is set by the ways in which it opens upon them or, alternatively, they are open to it. It must be remembered that in this way of conceiving the world, it is not to be simply identified with the entities that are *within* it—planets and atoms, oranges and houses. "World" is simply the fact of presence, of there being something there; and the ontological character of the world as a space of openness is thus linked with that of human being itself. This makes a good deal less strange the idea that the kind of entity a human being is finds expression in the kind of world it is in. What it comes to is that the being—the "What" of the things that are within the world—is disclosed (and thus "known") only to a being that itself opens on the world and that a certain modality of this openness corresponds to a character of what is disclosed. In this sense the character of that entity must be set largely by the openness to it of the entities that are within the world. Once again, it must be emphasized that this does not mean that the ontological type of the latter is the same as that of the entities in the world that are disclosed to it. The very fact that human beings are the entities to which this disclosure can be made as it cannot be, typically, to those that are disclosed constitutes a decisive ontological difference between them.

If in the light of these considerations we try to form a concept of human being that acknowledges and incorporates the fact of being-in-the-world, there will obviously be a question about how this transcendence-cum-presence of human being is to be conceived in relation to what we learn about ourselves through the good offices of the natural sciences, including the way our brains function. The difficulty here is that of thinking together the fact of transcendence and the neurophysiological account of the processes in the brain and the central nervous system that somehow make perception and other mental functions possible. It has already been noted that this difficulty is normally detoured around by describing the neural processes in question in language that simply enriches the properly scientific terminology of physics and chemistry with borrowings from the language of everyday life (and thus of the natural attitude) in which the presence of things outside the body is already presupposed. When we set aside this circular procedure, however, we come face to face with the fact that nothing about neural processes as described in the language of the physical sciences remotely suggests that they have any connection with the presence of entities or that they are necessary conditions for being-in-the-world. The blunt fact is thus that we simply do not know how this transcendence is effected. Even so, we can hardly ignore the countless empirical conjunctions of neural events and the presence and non-presence of things in the world. Everything we know indicates that in the absence of this brain activity there would be no transcendence and so we have to acknowledge the fact that it constitutes a necessary condition for what we are doing as whole human beings. I speak here of a necessary rather than of a sufficient condition because it remains impenetrably obscure how anything that happens in our bodies could make something outside, or for that matter inside, our bodies present. But if we have no basis for treating them as sufficient conditions of presence, events that occur in us evidently do, as just stated, play a role in connection with presence. Although we are quite unable to explain how they do so, our present understanding of ourselves as human beings has to be expressed by saying that we can define certain necessary conditions for presence but not any sufficient conditions. In this respect, our understanding of ourselves remains mired in paradox.

Ideally, philosophy would find a way of resolving that paradox, but it has been no more capable of doing that than natural science has been. The best it can do, in the face of the widespread indifference to these issues on the part of scientists, is to drive home the fact that there is such a paradox. Beyond that, if under these circumstances a philosophical concept of human being is to be advanced, it must be one in which the fact of transcendence is assigned a central position instead of being tacitly assumed. Such a concept will, therefore, have to acknowledge a gap in the story it presents, not between mind and body or between the physical and the mental, but rather between certain facts about human being as a whole—most notably, its character as transcendence—and the scientific account of the operation of the neural subsystems that evidently make it possible for these facts to obtain. Implicit in this notion of human being is the claim that the description of a human being as being-in-the-world applies to it as a whole and that by virtue of this fact the transcendence implicit in being-in-the-world is profoundly different from the circulation of the blood or the digestive process or even the functioning of the brain. There is no way in which we can assign a locale within our bodies to transcendence as being-in-the-world; and this is probably what makes it so easy to fail to recognize that the being-there of things is a fact about us as human beings. What I am proposing, then, is that, for purposes of forming a concept of ourselves, transcendence, with its correlative presence, be treated as primitive and placed at the center of that concept.

No such proposal is likely to pass unchallenged. For one thing, it may seem as though any such conception of the presence to a human being of things in the world must split this supposedly unitary entity once again and in a way that may be as hard to accept as the split that dualism effects. That, at any rate, might seem to be what any account of human being does when, like the present one, it seems to stretch it out over the gap separating it from the objects around it without any connecting tissue that would somehow make it one whole entity to which the term "unitary" could properly apply. I think it can be shown, however, that the sense of incongruity that informs this kind of objection really just reflects the conviction that human beings must, willy-nilly, conform to the ontological type of the entities around them so that they can be incorporated into nature without remainder. But to make this assumption is to endorse the idea that, in human being as in nature, everything is on either one side or the other of the lines that separate one entity from another. And that assumption makes it impossible to accept the most fundamental fact about human beings. This is the fact that human beings have a world, not just as a kind of neighbor they can drop in on occasionally, but as an essential complement of their own being and as, in fact, the locus of their lives.

Another likely objection here is the fact that transcendence is characterized by many of the same features that are familiar to us in connection with the concept of consciousness. The latter is the best approximation to that of transcendence that most of us can manage; and consciousness is plainly dependent on all sorts of contingent conditions and can be knocked out of commission by an anaesthetic or a sudden drop in blood pressure. It is true that the philosophical concept of consciousness is that of an intramental awareness of what is being represented in the mind—a representation of a representation, as Kant called it. As such, it belongs to the dualistic picture of a mind receiving input from a world that is external to it; and the on-and-off character of this "light of the soul" is thus attributable to the vicissitudes peculiar to a certain kind of entity rather than to anything that involves the world as such. Even so, it may well be asked whether the concept I am developing will not be susceptible to the same kinds of disturbances that affect consciousness. If so, the implications of this fact would be unsettling. Since the concept of presence is arrived at by dispensing with all mentalistic intermediaries, there would be no way of confining such disturbances to the inner state of the person so affected; and so it would follow that on any such view things in the world would be oscillating between presence and non-presence as this person's blood pressure rises and falls.

In trying to meet this kind of objection, we need to be reminded of the background beliefs that make it seem so formidable. We tend to think that if transcendence is to be what defines human being, its centrality should be matched by a stability and permanence that are in sharp contrast to the flickering awareness by which our lives are so often characterized. Conceivably, there may be beings whose being-in-the-world is not interrupted by sleep or by somatic disturbances of one kind or another; but we are not among them. Nevertheless, there is a sense in which the expectation just described *is* satisfied even in our case. Once presence is introduced into a life, even in the off-and-on manner that is ours, it governs the way that life has to be understood overall. In other words, the interruptions themselves to which it may be subject have to be understood in the terms appropriate to a being that transcends itself toward things in the world—that is, as hiatuses that have to be bridged over by postulating events that would have been disclosed if the hiatuses had not occurred. What this comes to is that presence introduces truth (and with it, of course, error) into our lives in a way that no other element in our nature does. It follows that the insecure status of presence in any human life cannot justify treating it as though it were simply another sometime "property" of human being and, as such, on the same footing as the others the natural sciences study.

This line of thought can be developed a little further. If presence has this special status in spite of its vulnerability to interruptions and disturbances of various kinds, then it may be possible to conceive a way in which it can be associated with the fact of brain function, not as its product but as its prior enabling condition. The idea would be that at ground zero a human being is an individuated locus of presence and that as such it is in the world and not simply a part of nature. As functions of such an entity, the electrical and chemical processes in our nervous systems could find their place and do their work within a life constituted by its opening on the world, although they would function in a way that we do not (and may never) understand. Conceived in this way, a human being would be a particular corporeal entity and, as we will see, an active one; but as a locus of presence and thus of truth it would be in the world and, as such, in nature only in a secondary, though vitally important, sense. If, as has been suggested, there is something inherent in human being but still unknown to science that makes it possible for us to inhabit a milieu of presence although it cannot bring such a milieu into being, then, to account for the interruptions to which our lives are subject, it must be assumed to function intermittently and sometimes imperfectly. But in that case, it would be possible, at least in principle, to account for the on-again, off-again character of our being-in-the-world without drawing either the substantive state of the world or presence as such into the little accidents that befall us.

IV

In this section an attempt will be made to amplify the conception of human being as transcendence that has just been proposed. On the basis of the arguments developed in the preceding section it will be assumed that transcendence as a holistic character of human being can best be understood in terms of the ways in which things in the world are disclosed to it. This means that what characterizes human being as a whole will be the modes of disclosure that correspond to significant articulations of the world upon which it opens.

An example that has already been used may help us to understand the way something about the world can show something about the distinctive character of human being. It is a familiar fact that in perception things present themselves to us in a certain orientation so that we see them from a certain angle and never see an object from all sides at one time. What is offered to us instead is a perspectival aspect—an *Abschattung*—of the object we perceive. In thought, we abstract from this perspectival character of our perceptual contact with things in the world and try to form concepts of them in which our position vis-à-vis those objects plays no role. Since, for this very reason, thought is held to be more "objective" than perception, it is tempting to conclude that it must also express the true character of our relation to those objects. It might even be suggested that, as so conceived, this relation is modelled on that of God to the world he created. But, like it or not, we do not stand outside the world as God is supposed to do and our knowledge of it is essentially mediated by perception as his is not. And what the perspectival character of perception shows is that we are in the world with the objects we perceive. Our position in that world determines the way in which something in the world is present to us. If this were not the case, it is hard to see why our relation to them should be perspectival. In this way, a fact about the world tells us something important about ourselves.

In order to develop this idea further, we have to be willing to embrace a conception of our opening upon the world that is very different from and far richer than our usual concept of "experience." Experience has typically been thought of as a kind of mosaic in which "ideas" (in Locke's language) or "impressions" and "ideas" (in Hume's) take up positions in a mental space, almost as though they were pieces occupying the place assigned to them in a jigsaw puzzle. The first modification that has to be made in that picture would recognize that these "pieces" are not ideas or representations that are in us; they are things and, instead of being the objects of a searchlight consciousness that plays over them, they are present to someone. Beyond that, the temporal character of the states of affairs that are presented to us also needs to be acknowledged. We are familiar with psychological ways of rendering such matters in terms of contrasts among perception, memory, and expectation; but an ontological approach in terms of the presence of the entities in question has to be cast in very different terms. This means, for example, that facts such as that a certain object has just appeared in a place where there was nothing like it before, or that day is about to turn into night, have to find a place in the way presence itself is conceived. It follows that there is a modality of presence that can only be called presence-in-absence and that what is not or is no longer the case as well as what will or may be about to be the case have to be accommodated in Presence as so understood.[7] This does not mean that past and future are to be given the status of actualities so that they can count as "real." Instead, what is actual is to be understood as "having been" or, for that matter, as "not having been"—in any case, as what follows upon whatever may have preceded it. What is about to happen—a familiar instance would be a butterfly's emerging from the pupa—also has its place in this kind of temporality. Another way of putting this would be to say that no actual "experience" is just an undated freeze-frame that is abstracted from the flow of events and then replaced by another such. It is itself temporal through and through; and in its Presence to us this temporality has to be recognized in the full complexity of the pastness and futurity that are the horizons of what is here and now. It is also clear that, in the case of time, orientation plays a role just as it did in the perception of things around us; and the perspectival character of our experience of sequences of events in time is the best evidence for our being ourselves in time as we are in space. In this case, too, we seem to want to do our best to conceive time as though, in our capacity as knowers and thinkers, we were not in it ourselves; and we do so in the interest of the kind of objectivity to which we aspire.

This is not the only way in which the actuality of the elements of "experience" has to be qualified. Over and above the contrast between what is the case and what has been or will be, there is also the contrast between what is the case and what might be or could be. Our understanding of the things with which we deal is not confined to their present actuality or indeed to their ascertained past or predictable future states. A stone is understood as something that can be picked up and thrown unless it is too large and heavy. And even if it is large and heavy, it is still something that one could *try* to move or to lift. Equally important is the fact that this stone can also remain just as it is. There is, of course, no guarantee that we will succeed in any of these efforts that we undertake although such efforts usually bring about at least some change that would not otherwise have been produced in the position or state of the object to which it is directed.

All of this applies not just to this or that stone, but to all kinds of things around us and it is something we take for granted in our dealings with them. The wind, for example, is something to catch in our sails or to make drive a mill. In this sense, possibility—what could be the case—is an uneliminable ingredient in our world and its counterpart is possibility as a constitutive element in human being. This is what is expressed by saying not just that this stone can be moved, but that I can (try to) move it. Possibility in this sense is arguably bound up with the fact of presence. Instead of simply acting on some feature of our being in the way fire makes wood burn, things in the world, by their presence to us, offer a range of possibilities that are realized in the responses we can make to them. Sometimes these possibilities are very restricted and sometimes they are indefinitely numerous; but in both cases it is as though a unique kind of hiatus had opened up in the causal chain and permitted alternative possibilities to present themselves. Presence would, accordingly, thus be linked with a degree of freedom in the conduct of our lives.

All of this has a special relevance to the understanding we have of the causal relations in which things in the world stand to one another. As David Hume famously demonstrated, it is not as though those relations were reliably evident simply to empirical observation. In order for these entities to be present to us as having these causal powers, it must be possible to vary the context in which they are encountered and this means to set them in the context of what they would do or be if some other event were to occur. But a human being can do that only if it is capable of acting in such a way as to bring about some such event in circumstances in which it would not otherwise occur. What this comes to is that things are understood not just in terms of what they presently are or predictably will be, but also in terms of what they can do—that is, of what they would do in certain circumstances that may not arise but can be brought about. Because human beings understand the things they deal with in these terms, they can be said to understand them in a context of possibility—that of what could or might possibly be done with or to or for these things. They can be grasped or moved or eaten and as such they are primarily things in use. But this entails that they form part of the field of a being that itself has to be thought of in terms of what it *can* (try to) do.

In the standard scientific concept of nature, by contrast, all such processes are typically reduced to their pure actuality and stripped of anything that would have to be rendered by the use of the word "can." If this requirement were consistently adhered to, as it mostly is not, even the experiments we make to determine how different kinds of events are connected with one another would have to be treated, not as methods for determining what something *can* do, but as events on their own in their own causal order. In this way a Megarian world of pure actuality would be constituted that consists of all the things there are, arranged in their sequences along a world line. Such a world would be described by "eternal sentences" that abstract from pastness and futurity as "subjective." By contrast, the argument being made here is that even if the natural world is to be conceived on this model, it could never be "known" to be such if the entity by which it is known were frozen into the same kind of tenseless actuality and denied any linkage to what is possible. This is what is implicit in the preceding discussion of "can"—the word that expresses a fact about how one thing would affect another even if there were no occasion for the one to act upon the other. This is also what causal knowledge turns on and in the absence of such knowledge there could only be sequences of events without any basis for the claim that one event produces another.

What has been said thus far about the way the character of the world expresses the character of human being comes down to this: human beings are in the world in the mode of disclosure and they are active in a sense that involves possibility. The two poles of a human life are thus truth as the openness to us of things in the world and possibility as the enabling condition for action. Together with the equally essential fact that for human beings outcomes matter differentially, these generate a purposive ordering of what happens in the world, so that not just artifacts but even natural processes are understood in terms of the consequences they might produce as these affect the interests of the human beings in question. Action can then be organized either to head off adverse effects that will predictably occur unless they are prevented or to bring about desirable outcomes that some turn of events makes possible. Nothing of this kind can occur unless the actual and typically non-purposive patterns of events around us have been identified and utilized in a context established by an understanding of the world in terms of what can happen and what can be done.

There is still another respect in which something about the world defines a significant feature of human being. In the world, there are other human beings—entities that are in the world as we are and are interested and active in the same generic sense as we are. These beings are identified, not just by their physical similarities to us, but by the fact that they perceive and work with the same objects as we do and by the purposive character of their activities in the world generally. As has already been explained, this becomes the basis for a communicative relationship among human beings that is marked by both conflict and cooperation. The network so formed of ties to other human beings, not only those who are with us now but those who have preceded and will follow us as well, is the matrix in which individual personality and character are formed. It is implicitly assumed by all forms of inquiry into such matters, even when some behavioristic or quasi-behavioristic philosophy is officially espoused by those who conduct them. This is to say that human life is, in a very broad sense of the term, moral and in this respect, too, it generates relationships for which the natural world offers no models.

In light of all these forms of transcendence by which it is characterized, the best overall description of human being would be to say that it lives beyond itself as an organism. It is irreversibly (and in its own interest) drawn into what is going on around it and most especially into relationships with other human beings. It is not a neutral spectator vis-à-vis the outcomes it can anticipate for what is occurring in its world as these impinge on its interests. One could even say that the future as what is about to happen is the dominant temporal modality of our lives and as such "the place" where we live much of the time. This is a future that is understood in terms of lines of concern that come out of our pasts and the expectations that seem justified to us on the basis of past experience, our own and that of others. All these hopes and expectations are bound up with lives other than our own as are our deepest disappointments and our most intense moments of happiness.

One could amplify this description of human being by saying that its character is such as to make the natural entities by which it is surrounded the scene of its life. This is in fact the import of the concept of world as that which a human being is in. In a freely metaphorical mode, one could say that each of us and all of us together weave a web of meaning around the place and date of our lives and that the natural environment into which we are born is overarched by the architecture of those lives. In these circumstances it is not hard to understand why each of us is for himself the center. That is often condemned as parochial and vain; but to this one might reply by asking what other kind of center there could be. Since it is precisely within this network of meaning we weave that the being of things in the world is disclosed as well, perhaps this claim should not be dismissed quite as contemptuously as it is by those for whom galaxies and big bangs are the ultimate reality.

One further observation suggests itself in this connection. In Chapter 3, I spoke of a disjoining of perception from its object and of the need to revise our thinking about perception so as to re-establish the proper nexus between the one and the other. The vice in the kind of thinking about human being that leads to hard naturalism and to which attention has been drawn in this section can be viewed as this same disjoining writ large and no longer confined to perception or any particular department of "mental" functioning. It is as though the self-confidence of naturalistic theory had risen to the point where it no longer acknowledges that it owes anything to anyone and least of all to the humble modalities of presence that make the world available to it in the first place. All those modalities have been treated in the same manner as all the other functions of human being; and this means that they have been incorporated into the object domain and its processes over which that theory presides. The opposing thesis of this book is that theory must not be allowed to devour, not so much its children in the manner of Chronos, as its parent. It is surely not entirely fanciful to conceive the milieu of presence that is the world as a receptacle in the Platonic sense and thus the womb of theory. But in that case there is surely a real hubris implicit in such claims on the part of theory; and one day, one cannot help thinking, the progeny it begets will be called to account for the peculiar circumstances of their birth.

V

In Chapter 1 it was pointed out that although human beings have traditionally been at least partially taken out of nature, non-human animals have been treated very differently. These "brutes," as they were so often called, might have souls; but they were of a distinctly lower order than those of human beings. According to Aristotle, animals had vegetative souls since they took nourishment and grew and they also had sensitive souls since, unlike plants, they were capable of perception; but they did not have rational souls and so they could not in any proper sense think or reason as human beings can. Other philosophers were willing to push the contrast between human beings and other animals even farther; and some, like Descartes, claimed that animals were simply cunning automata and had no souls at all. In a way, contemporary naturalism could be characterized as an extension of Descartes' idea by which it is applied not just to non-human animals but to human beings as well.

I have been arguing that human being is defined by transcendence and that may well sound as though I were claiming that the association of that concept with human being were unique. That was not, however, my intention and I must now face up to the question about how much of the apparatus associated with this concept can be attributed to non-human animals. It was pointed out in Chapter 1 that anti-naturalists have typically insisted on placing such animals in the domain of nature and have wanted to draw as deep a distinction between them and human beings as they possibly could. This would mean that the distinction between nature and what is not nature would also be a distinction among kinds of living things. It is at best unclear what provision would be made under these circumstances for the contrast between nature and the world as a milieu of presence. In any case, that is not the course I propose to take. As far as I can see, there is no good reason for denying that animals other than humans inhabit a milieu of presence. The dog that chases the ball I throw out sees the same ball that I see and, if it disappears in the undergrowth, searches for it just as I would do. Again, I am present to the dog and it is present to me; it recognizes me as its master and awaits me on my return home in a way that certainly suggests a working familiarity with the distinctions between past and present and present and future. In all of these respects there seems to be no basis for the presumption that presence is confined to human beings.

HUMAN BEING AS THE PLACE OF TRUTH

It does not follow that the milieu of presence—the world—in which animals live is indistinguishable from the world human beings live in. There is every reason to think that the lives of animals are, in Heidegger's phrase, *weltarm*—"world-poor"—in the sense of being impoverished in respect of the features of things in the world that they are able to register. The things that are present to me and to my dog are in many important respects the same, but there are, nevertheless, differences in the way they are present. In a general way, one can say that although the concrete and the immediate loom large in the lives of both animals and human beings, human beings are able, admittedly in varying degrees, to make a place in their way of having a world for things and facts that are removed from their present situation and their immediate needs. This is something that animals are largely unable to do; and there can be little doubt that the fact that we have highly developed languages, as other animals do not, plays an important role in this difference. What language as the presence-in-absence of things in the world does is to enable us to place those things and facts and events in our world in a way that reflects the experience of many people other than ourselves as well as a large measure of independence from any practical utility that such information may have for any given individual at any particular time. What I am suggesting is that, lacking language and the distance from the concrete here-and-now that it makes possible, animals are bound to the immediate situation in which they find themselves in a way that we are not. It is one thing to be able to recognize and expect this or that person or thing; it is quite another to be able to contrast present with future or present with past across the board. Because they cannot do this, animals have to live in the present and with the objects in that situation to which their needs relate; and they are unable to understand the world as a stable yet changing array of things and animals and people.[8]

This point has an important corollary. Because the things animals are familiar with and pick out of their environments are the ones that relate to their basic needs, the identities of those things will be pretty well fixed in that one way. If human beings have sometimes thought of things in the world as having essences—properties that defined them in a special way and from which they could not be separated—one could say that that kind of fixed identity is realized in the world of animals. This means that it will be very difficult and often impossible to bring a given entity that has been implicitly classified in one way under any other "description." As a result, the possibility, that is so central to human culture in all its forms, of tinkering with such identifications and seeing affinities and similarities that are not laid down in the official code must remain largely closed to animals.

Trying to understand what kind of world the higher animals live in is a difficult business at best; but we may be aided in our effort to do so by a parallel that suggests itself with human beings who have suffered serious brain injuries.[9] Such persons have acquired language, but they are nevertheless bound, in varying degrees, to what is concrete and immediate to a much greater degree than is a normal human being. For them, references to things in their situation by means of terms involving relations that reach beyond that situation become increasingly unavailable. A more general way of making the distinction between the ways in which animals and humans are in the world would make this distinction turn on the role of self-consciousness in the one case and the other. "Self-consciousness" is a notion that belongs to the vocabulary of mind and as such has no obvious right to be invoked here. It can, however, be translated into the language of presence where it would be rendered as the reflexive relation to self of which an entity constituted by presence is capable. In other words, a human being is given to itself as the transcendent being it is. Sometimes animals have been said to lack not just self-consciousness in this sense of reflexivity but every other imaginable form thereof as well. But if this is supposed to mean that they cannot distinguish between, for example, a threat to one of their congeners and one to themselves, it cannot be right. The status of self-consciousness in the lives of animals is presumably like that of their understanding of temporal differences. Both are evinced to some degree in the way animals respond to certain kinds of situations, but neither can be recognized as what it is in some degree of independence from those situations. The self-consciousness animals radically lack is, accordingly, an awareness of themselves as subjects—beings for whom things are there.[10] It is often difficult for human beings to get such a sense of themselves; and this book has recorded the distortions to which that sense is vulnerable even when it is acquired.

Kierkegaard referred to human beings as "universal particulars"—particular because each of us is just the one person he or she is and universal because we have a world and are familiar with an indefinitely extensive array of entities other than ourselves which we bring under a variety of headings on the basis of their similarities and dissimilarities to one another. In some cases, the element of universality can predominate to such an extent that, as Kierkegaard also noted, the element of particularity is at least neglected if not positively endangered in the understanding such persons have of themselves. Nothing like this can properly be attributed to non-human animals. To the extent that human beings are able to achieve a freer way of having a world, they can also be said to be familiar with the fact of meaning. In this sense, "meaning" would just be what something/anything is, considered independently of any actual occasion of its having that character. There need be no Platonic overtones—nothing implying the independent subsistence of abstract entities—in an acknowledgment of such a capability on our part. I can, for example, *pace* Berkeley, understand that any three-sided object is a triangle without being committed thereby to the existence of a three-sided object that is neither scalene nor equilateral nor any other specific kind of triangle. What is involved is simply a more complex way of living in the presence of a world of things with their own multiple characters which we are able to distinguish from the particulars that bear them.

If in the light of all these considerations we ask whether non-human animals belong to nature or to the world, the answer has to be that in a fundamental ontological sense they belong to the world. Within the world, however, one has to recognize a distinction between degrees of freedom in the relation in which living creatures stand to their natural environment. That distinction separates human beings from other animals in a radical way; and it is likely that that difference will continue to weigh more heavily in our sense of what belongs where than does the fact that animals are not, in an ontological sense, things.[11]

CONCLUSION

The principal conclusion about hard naturalism that emerges from this study is that it manages to keep going only by constant unavowed borrowings from both dualism and the natural attitude. In its authentic unadulterated version it is simply too obviously in conflict with itself and with its own claims about what human beings are like to be credible. Its authority is wholly derivative from that of the natural sciences of which it claims to be the only valid philosophical interpretation. In return, however, it does these sciences a considerable disservice by associating them with an untenable philosophical thesis that also distorts out of all recognition the most fundamental facts about human nature.

This conclusion was supplemented by a sketch in Chapter 6 of a more satisfactory conception of human being. The status that is assigned to presence within that conception is one that makes it, not just a by-product of neural functioning, but the ontological fact on which all science and all of human life and culture are founded. As such, it is not susceptible of being drawn into the circle of causal explanation of the natural sciences. Presence and transcendence simply do not belong to the domain of fact in which such explanations are properly sought and they do not belong there because they are what makes such inquiries possible.

The issue with which philosophy has to deal in the aftermath of this conclusion is whether it is possible to go forward from it and to think together world and nature, presence and neural functioning in some way that is more satisfactory than those that are now available. Somehow, we cannot help feeling, it should be possible to conceive a wider reality that makes a place for both the one and the other and does so in a way that would enable us to understand a deeper linkage between them than the account given here provides. Even to suggest that this is needed reminds us, of course, of the grandiose attempt that Hegel made to construct just such a dialectical unity of thought and things. That precedent, for all its genuinely impressive character, must make us skeptical about the prospects of any such undertaking. It may be, after all, that the kind of account of world and nature that has been sketched here represents the most that our radical finitude permits us to

CONCLUSION

affirm, with any justified confidence, about this subject. Nevertheless, the effort to work something out at this highest level of philosophical thought will undoubtedly continue and there is no reason not to hope that it may produce something of value.

Whether or not anything more ambitious is possible, there can be little doubt that the present account will seem deficient to many. It makes no attempt to supplement the bare phenomenological facts it recounts with any other kind of story. In some other treatments of these themes, there has been a marked tendency to do that. Very often the underlying agenda in such cases is not just a desire to set limits to the epistemic authority of the natural sciences, but to make a case for some other kind of knowledge and, most notably, for religious ideas of one kind or another. It has long been assumed that religion and science are locked in a zero-sum game and that if science loses points, these must accrue to religion. I have done my best to avoid any such assumption and so the conclusions I have reached do not support claims that have been made by any large ideological rival of natural science on the current cultural scene.

It is hardly surprising that we should want to have this sort of amplification of the account of human being that has been given here. The great fact about the life economy of human beings is that it has a pervasively purposive character. To the extent possible, everything we do is done for a reason—not necessarily a good one—and to achieve some goal that is judged to be preferable to other possible outcomes. Even when we reach the margins of human life, we want to be able to ask why we are here and why there should be any such creatures as we are. Plainly, any answers that could be given to these questions would have to invoke the idea of purpose and would attempt to show that the purposive ordering of our lives is embedded in a larger system of purposes that would be attributable, presumably, to our creator. To cite just one example, it has been claimed that presence has the character of a gift that has been made to human beings. Such a gift would open up a relationship with a supernal being of some kind; and such a relationship, in turn, would presumably constitute the ultimate meaning of our lives.

For the time being, at least, we will have to settle for a much more modest set of conclusions. There is a larger meaning that attaches to the conclusions about human life that have been proposed here; but it is of a quite different order from the kind just described. This is not to say that these conclusions necessarily exclude every religious interpretation of our lives; they are just not designed to favor one. Instead, what they declare is simply that there is a world and not just nature and that there is no point in trying to collapse the former into the latter. This may well sound a little flat and inconsequential, but it should not. What it expresses is a new way of conceiving the contrast between the natural order and something that transcends it and of doing so without sealing the one off from the other as happens when the concept of the mind or the soul is invoked for this purpose. On this account, the basis for this contrast is the presence of the entities that make up nature to an entity that is itself in the world, but in it in a way that is quite different from the mode of being of most of the entities that are present to it. Because presence resists conceptualization in the terms made available by the scientific worldview, naturalism as the philosophical formulation of that position cannot accommodate it. The plain implication of all this is that, in equating nature and world, naturalism has got things wrong; and that is the conclusion this book undertook to establish.

The relations between nature and world are complex and badly need sorting out. What we are initially and mainly familiar with is the world although we often confuse it with nature. We arrive at the scientific concept of nature only by stripping away the features of the world that betray the fact that it is the domicile of human being and, as such, lends itself to an ordering that has a purposive character. This kind of reduction of the world to certain of its properties and relations in favor of others has taken place principally in Western thought. It is worth considering that this reduction might very well never have occurred and that in that event the familiar entities of everyday life would not have been displaced by others that originate in theory. I say this, not in the spirit of expressing regret at the discovery of entities outside the common-sense framework, but to suggest that things might have gone differently and that the continuity between common sense and theoretical science might have been better understood had it not been for the way that framework itself was treated.

What I have in mind is the fact that in Western philosophy our way of being in the world came to be understood in dualistic and representational terms. As was explained in Chapter 2, the "sensible appearances" that were supposed to mediate our relation to the *res verae* in the world were assigned to the mind as the domain of the subjective—the not really real—and a domain of extramental entities was postulated behind those appearances. The phasing-out of this kind of mediation was also described in Chapter 2; and this led to the claim of naturalism that only those natural entities, sometimes known as things-in-themselves, really exist. In this way the whole phenomenon of the world as what we are in was missed, first by its being misconceived as a kind of inner landscape and then by its being abolished altogether because no such conception could be squared with the criteria of reality employed by the sciences.

The thesis that has been proposed in this book attempts to reverse both of these profoundly misleading judgments. It resists the internalization of our perceptual life and argues instead that the entities we perceive provide the starting point for all the subsequent analyses of their constitution out of which the theoretical entities of the various sciences emerge. Because they are what we perceive, we are, *ab initio*, really and irreversibly out in the world as we would not be if we were dealing only with mental representations of those entities. At the same time, it is acknowledged that this fact—the presence to us of entities that are not in our minds—cannot be explained by the methods of the natural sciences. If it is rejected on those grounds, however, we will never find a better reason to affirm the existence of things other than ourselves.

The main significance of the fact that we live in the world and not in nature is that the destructive impact on our understanding of ourselves of the scientific thesis that we are denizens of nature is much reduced. It is still true that, like other animals, we are born and we die and in between we are all vulnerable to "the thousand natural shocks that flesh is heir to." Even these are different, however, in the way they figure in our lives since there is every reason to think that we live with such facts about ourselves through anticipation and recall otherwise than other animals do. What is even more important is that the inhibitions imposed by the conception of nature as our habitat, on the characterization of a whole range of human functions, are in effect suspended. Beings that live with one another, as we do, in a condition of transcendence—transcendence of the organism that each of us is and toward things in the world and one another—cannot be understood simply in terms of the causal regularities governing the processes that take place in those organisms or in the natural environment around them. This is not to suggest that our life situations do not remain hedged about by all kinds of constraints, both material and epistemic. What it does imply is that among them there need be no obligation to make our understanding of ourselves conform to the explanatory norms that govern the work of the natural sciences. That is the obligation that naturalism seeks to impose and, as should be clear by now, it thereby distorts the very inquiries it claims to set on the royal road to the truth.

NOTES

PREFACE

1 I gratefully acknowledge the assistance I have received in connection with this project from Professor Alastair Hannay who read a penultimate draft of this book and offered many valuable comments and suggestions.
2 E.O. Wilson, *Consilience: The Unity of Knowledge* (New York: Vintage Books, 1999).
3 Wilson describes the work of John Rawls as traveling the "transcendental road." It does not seem to occur to him that the normative character of the themes dealt with in *A Theory of Justice* require a different mode of treatment from that of empirical studies of human conduct.

1 NATURALISM IN HISTORICAL PERSPECTIVE

1 John Stuart Mill, "Nature" in "Three Essays on Religion" in Marshall Cohen, ed., *The Philosophy of John Stuart Mill: Ethical, Political, Religious* (New York: Modern Library, 1961), pp. 445–88.
2 The original formulation of this idea was in T.W. Adorno and Max Horkheimer, *Dialectic of Enlightenment*, translated by John Cumming (London: Verso, 1972).

2 NATURALISM, DUALISM, AND THE NATURAL ATTITUDE

1 I take this phrase from E.A. Burtt, *The Metaphysics of Modern Physical Science* (London: Routledge & Kegan Paul, 1932).
2 The classic critique of such a conception of perception is to be found in Maurice Merleau-Ponty, *Phenomenology of Perception*, translated by Colin Smith (London: Routledge & Kegan Paul, 1962), Introduction.
3 A more detailed discussion of the argument from error can be found in my book, *Heidegger and the Philosophy of Mind* (New Haven: Yale University Press, 1987), Chapter 2.
4 A curious thesis concerning the relation of physical science to the existence of human beings has been propounded under the title of the "anthropic principle." Its claim is that if certain physical constants had been slightly different—the ratio of the number of photons to protons that permits the formation of carbon atoms, for example—human life would not have been possible and the science of physics would never have come into being. What implications, if any, can be drawn from this fact is, however, far from clear. See J.D. Barrow and F.J. Tipler, *The Anthropic Cosmological Principle* (Oxford: The Clarendon Press, 1986).

NOTES

5 This is a quotation from one Andrei Linde, described as a "Stanford University cosmologist," that appeared in the "Science Times" section of the *New York Times*, 18 June, 1998, p.B12.

3 THE REJECTION OF THE GIVEN AND THE ECLIPSE OF PRESENCE

1 Among the very few philosophers who have addressed this question, one must give honorable mention to Maurice Merleau-Ponty and to his conception of our bodies as *la chair* (flesh) rather than simply the matter of the natural sciences. In the background of all such discussions is the doctrine of Aristotle in *De Anima* of the soul as having no "nature" of its own that would interfere with the reception of the forms of the entities it comes to know.
2 The work of Wilfrid Sellars has perhaps been more influential than that of anyone else in bringing down "the myth of the given" in American philosophy, especially in his famous essay, "Empiricism and the Philosophy of Mind." How he would deal with the concept of observation in the absence of any conception of the given, however reconstructed, remains unclear.
3 This phrase stems from a well-known article by Thomas Nagel, "What Is it Like To Be a Bat?" *Philosophical Review*, 83 (1974), pp. 435–50; and it has been taken up by David Chalmers in his recent book, *The Conscious Mind* (New York: Oxford University Press, 1996).
4 I wish I knew who said this, but I have been unable to find out who it was.
5 I have in mind here the work of John McDowell and his book, *Mind and World* (Cambridge, Mass.: Harvard University Press, 1992).

4 THE SUBSTITUTION OF LANGUAGE FOR PRESENCE

1 The reference here is to Daniel Dennett and his article, "Are Dreams Experiences?" *Philosophical Review*, 85 (1976), pp. 151–71.
2 On this point, see Hans-Georg Gadamer, *Wahrheit und Methode* (Tubingen: J.C.B. Mohr/Paul Siebeck, 1965), Part III.
3 I have tried to develop this idea further in my book, *Heidegger and the Ground of Ethics: A Study of Mitsein* (New York: Cambridge University Press, 1998).
4 This idea is developed in Maurice Merleau-Ponty, *The Visible and the Invisible*, translated by Alphonso Lingis (Evanston, Ill.: Northwestern University Press, 1968).
5 The bizarre idea that a truth-claim that is made without relativizing it to some individual or group is a form of personal aggression is peculiar to our own time and will not be considered here.

5 WHAT DOES THE BRAIN DO?

1 An excellent account of the way various body parts and organs have been identified as the locus of mental functions is given in R.B. Onians, *The Origins of European Thought* (Cambridge: Cambridge University Press, 1951).
2 I take this remark by Wittgenstein from John Cook, *Wittgenstein's Metaphysics* (New York: Cambridge University Press, 1994), p. 209.
3 John Searle in his *The Rediscovery of Mind* (Cambridge, Mass.: MIT Press, 1993) seems to view this as a minor difficulty, but it is surely more than that.
4 The reference here is to Colin McGinn, *The Problem of Consciousness* (London: Oxford University Press, 1991).

5 As an example of this tendency I would cite a work on children's consciousness of other human beings, S. Baron-Cohen, *Mindblindness* (Cambridge, Mass.: MIT Press, 1995), in which it is simply assumed that this ability is to be accounted for by postulating the existence of a corresponding "device."
6 For a fuller account of this approach to the problem of error, see my *What Is a Human Being?* (New York: Cambridge University Press, 1985), Chapters 2 and 3.

6 HUMAN BEING AS THE PLACE OF TRUTH

1 The reference here is to a puzzle posed by Professor Searle in which a translation is made of certain passages in Chinese into another language by someone who does not know any Chinese or the other language and therefore has no idea of what the meaning of the passages he translates may be. He makes the translation by relying entirely on dictionaries and rulebooks that set forth the equivalences between Chinese expressions and others in the second language. See John Searle, "Minds, Brains, and Programs," *Behavioral and Brain Sciences* 3: 417–24.
2 I take this expression from the writings of Virginia Woolf who uses it in a context in which she has a character speak of occasions when these walls grow "thin" and "nothing is unabsorbed." All the paradoxes inherent in dualism are wonderfully, though unconsciously, captured in this statement. See Virginia Woolf, *The Waves* (London: The Hogarth Press, 1931), p. 224.
3 The expression one might equally well use here is the one devised by Martin Heidegger: "ek-sistent." It plays on the etymology of the Greek and Latin words for "exist" to form an expression that means something like "to go beyond" or "stand outside." In the interest of avoiding what may seem rather exotic, I have used the more familiar "transcend" although it has some connotations that may be misleading.
4 This ancient sense of the word *objectum* presumably explains the later use of the word "objective" in medieval philosophy to denote what is in the mind. This has puzzled many students because it is so directly opposed to the modern understanding of "objective" as what is really out there in the world. Descartes' use of the term supplies the effective link between its medieval and modern meanings. For him, "objective" applied only to those properties of things in the world that had passed the mind's test and been shown to be "clear and distinct" and thus in conformity with the requirements of reason. This new demand for "objectivity," in turn, has led to a certain tendency, noticeable above all in Heidegger's writings, to shy away from the word "object" (and its German cognate *Gegenstand*) on the grounds that it designates only what has been tidied up and generally put in good logical order by the mind as a condition of its being recognized as real. What is found to be objectionable in this is the implicit claim that the human mind is altogether sovereign in these matters and its criteria are thus the "measure of all things."
5 Martin Heidegger, *Gesamtausgabe*, II, 56–7 (Frankfurt am Main: V. Klostermann, 1987), p. 88.
6 In Western philosophy the *locus classicus* for an ontology of light is Plato's discussion in the *Republic*, Book 6, of what he calls "*ta epekeina tes ousias*"—what lies beyond being. Here the sun of the sensible world which corresponds to the idea of the good in the intelligible world is described as what binds together vision and its object.
7 The idea here is that Presence should serve as the overarching concept that expresses what is common to presence and presence-in-absence.

NOTES

8 Everything I say here about animals really has to do with the so-called higher animals. I simply don't know what to say about sea-slugs and paramecia.
9 I am drawing here on the work of Kurt Goldstein and especially his *Human Nature in the Light of Psychopathology* (Cambridge, Mass.: Harvard University Press, 1963).
10 There is a passage in the *Eighth Duino Elegy* in which Rilke describes animals as inhabiting a space that he calls "das Offene" and as doing so in a very different way from that of humans. They also perceive humans as being somehow not at home in the world as they are.
11 The question whether any animals can be said to have an implicit familiarity with being in the Heideggerian sense is not easy to deal with. They are certainly not able to isolate or "thematize" such features of their world as temporality or possibility and these are central modalities of being as Heidegger understands it. Although he does not appear to require this as a necessary condition for a familiarity with being, all he appears to be willing to grant animals is what he calls "access" (*Zugang*) to entities. Even so, one is tempted to ask how there can be "entities" in the absence of "being."

INDEX

animals: as modality of presence 101–4
Aristotle 71, 101
Austin, J.L. 49

being: our familiarity with 57; of things 40
belief: as disposition 75; and perception 74
Berkeley, George 13, 103
body, human: in perception 72, 34; as physical system 22
brain: and belief 32, 56; function of 65–83; in natural attitude 18; as seat of intellectual powers 66–7; as site of perception 18, 28–9, 32, 36–7; in a vat example 80
Butler, Joseph 44

causality: and agency 98–9; as external relation 23; and human being 105; in perception 38–9, 40; space of causes 42
cognitive science: and philosophy 65
color: discrimination of 32; as property 26–7; as subjective 13–14, 20
communication: among human beings 57–60
consciousness: and brain function 67–8; concept of 36–7, 94–5; in cosmic evolution 77; inexplicable by us 75–6; in Hegel 21–2; in naturalism 74; terminology of 2, 13

Democritus 6, 65
Descartes, René 6, 88, 101
Dewey, John 5, 6

dreams: accounts of 46
dualism, mind–body: as account of perception 31–2; and brain 42–4, 68, 75; critique of 15; in Husserl 17; and naturalism 15–16, 69–70, 80, 86; paradoxes of 29; and privacy 30–1; and realism 16; as response to anomalies 14–15; and scientific world-view 12–13, 14–15; as theory of mind 10, 11–12, 84, 86–7

Enlightenment, the: dialectic of 3–4
error: as affecting relations 88
evidence: and presence 30–1
experience: and brain function 67; and evidence 30–1; of pure quality 27–8; traditional concept of 24, 96–7

feeling: and perception 38
folk psychology: dualism as 68, 72
functionalism: as theory of mind 71

Gadamer, H.-G. 57
given, the: as ontological fact 19–20, 31; and presence 39; rejection of 24–31

Hegel, G.W.F. 21–2, 105
Heidegger, Martin ix, 17, 30, 48, 57, 89, 91, 92, 102
Hobbes, Thomas 6, 65
human being: contrast with intra-mundane things 93; as locus of presence 95–6; as place of truth 84–104; as reflexive 102–3; relational character of 86–90; and world 94–5

human beings: as being in the world 23; consciousness of 38; as "other" 99–101; role of, in neuroscience 72–5; scientists as 84–6; and truth 42–3
humanities: as conception of human being x, 7
Hume, David 13, 97, 98
Husserl, Edmund ix, 2, 6–17, 22, 49

idealism: and critiques of dualism 13
intentionality: of words 56
introspection: concept of 81; defects of 46

judgment: in perception 29

Kant, Immanuel 6, 87, 95
Kierkegaard, Søren 5, 103

language: and animals 102; as communication 57–60; as disclosure 53–7; "invisibility" of 57; and perception 29–30; 33–4, 49–50; as physical event 51; as presence in absence 102; role in contemporary philosophy 25; and states of affairs 52–3; as utterance 48, 71; worldly character of 48–9
light: presence conceived as 91–2
Locke, John 24

manifest image: and physical theory 20–1
Marx, Karl 77
materialism: conclusions about 105–7; and naturalism 6, 22, 65; and physicalism 51, 71, 80
Mead, G.H. 57
Merleau-Ponty, Maurice ix, 58
Mill, J.S. 1
mind: and brain 67, 69–70; and category of relation 88–9; concept of 2, 10–11, 12–13, 14–15, 29; and language 46–7; and perception 32; and presence 84, 88; as private 15; role in this inquiry 23; and sense data 27–8; skepticism about 13; as substance 14, 88; as transparent to itself 92

natural attitude: defense of 17–19; Husserl on 16–17; implicit in scientific world-view 69; and language 47, 48–9, 93; and "naive" realism 16; and naturalism 21–2, 74; in neuroscience 37; not a theory 19; in perception 34; and truth 80–1
naturalism: and belief 32–4; and brain 67–8, 71, 75–6; as critique of dualism 14–15, 22–3; development of ix, 3, 4–8; "hard" and "soft" 7–8, 66; and language 50–1, 54, 56; and mind ix, x; and natural attitude 21; and perception 31, 38–9; and presence 29, 58; and privacy 31; and states of affairs 52–3
nature: and animals 1–2, 101; concept of 1–4; and world 78, 84–5, 105–6
neurophysiology: mode of inquiry 72–5; and representations 80–1

objectivity: contrasted with subjectivity 11, 13–14; as elimination of possibility 97–8; non-oriented character of 96–7; and norms 62; and objecthood 90–1; objectification of self 50–1; and truth 42, 43; and world 60–1
ontology: of light 91–3; of naturalism 7, 22; of things 88–9; of words 48–9, 60
orientation: in life-world 78; perceptual 96

pain: as modality of presence 81–3
perception: as access to object domain 78–9; as behavior 34–5; and belief 33–4; as brain event 28–9, 36, 37, 48; causal character of 38–9; in contemporary philosophy 25; as disjoined from objects 100; as external relation 54; as fallible 11–12, 17–18, 24; and language 49–50; and naturalism 31; as relation 87–90; reports of 73–4; and theoretical entities 107; transmission theory of 68–70, 79–80, 91; visual 58
possibility: as modality of presence 97–9
Platonism 41, 101, 103
presence: and absence 48, 55, 88, 97, 102, 118; as affected by bodily states 94–5; and animals 101–4; and communication 57–60; conceived as relation 39–40, 191; and existence 25; inexplicable by natural science 76–7,

INDEX

106, 107–8; and language 29–30, 118; as light 91–3; as "milieu" 58–60, 85–6; and natural attitude 22; and naturalism 29, 31; noncausal character 105; and pain 81–3; as primitive concept 76–7; as replacing mind 84; as space of reasons 42; as temporal 96–7; as transcendence 56, 86; and truth 42–3
privacy: and dualism 30–1; and language 46
purpose: in human life 106–7

rationalism: its treatment of perception 25
realism: in Heidegger 17; "naive" 12, 16, 54
reference: of words 50, 55–6
relation: external 44, 54; and human being 86–91; internal 89; presence as 40, 88; syntactical 52
relativism: and scientific world view 62
religion: and science 106
representations: as contents of mind 10, 79, 80–1; in dualism 14, 18, 29; on "inner screen" 69–70; neural 28–9, 37, 72–3; and perception 12; as term in relations 89

Santayana, George 41
Sartre, Jean-Paul ix
science, natural: and manifest image 20–1; and mind 13; and naturalism 3–4, 8, 14–15, 66; as theory of world 20
scientific world-view: as denial of presence 76–7, 79–80; and dualism 14–15; and natural attitude 74, 107; and naturalism 8–9
Searle, John 85
self-consciousness: of animals 102–3; human 22, 36–7
sense-datum: concept of 25–8; dismissal of 28; in dualism 29; and judgment 29

Snow, C.P. x, 7
states of affairs: as expressing presence 51–3
Stoics 1
subject: animals as 103; and mind 11; new concept of 86; reality obscured 44; subjective–objective contrast 11, 14, 20, 23; as temporal 99; as "worldless" 14
substance: as foundational category 87

teleology: and brain function 71
temporality: of presence 97, 100; of animals 102
transcendence: and bodily conditions 93; and brain function 93; of human being 86, 94–5, 100, 105, 108; of language 56
truth: and brain function 80–1; and extensional logic 43–5; and gaps in mental life 95; and language 47; as normative 60–2; and presence 42–4; and states of affairs 51–3

values: and relativism 62–3

Weber, Max 4
Whitehead, A.N. 24
Wilson, E.O. x
Wittgenstein, Ludwig 6, 17, 67
words: and sounds 54–5; and things 55–6; and world 57
world: and animals 104; contrast with nature 85–6; genesis of 90; and language 92; as "milieu of presence" 92; not creature of theory 18, 19; as "objective" 60; passed over by Western philosophy 107; as shared 57; as space of reasons 42; and states of affairs 51–3; as web of meaning 100; as zone of openness 23, 40, 59; 84–5

Yeats, W.B. 4